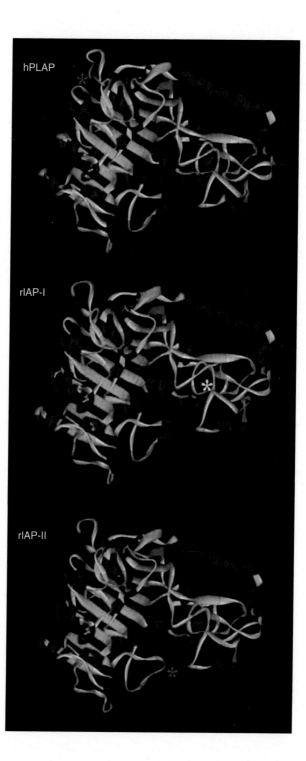

BIOCHEMISTRY FOR MEDICAL PROFESSIONALS

BIOCHEMISTRY FOR MEDICAL PROFESSIONALS

TSUGIKAZU KOMODA
Saitama Medical University School of Medicine, Saitama, Japan

TOSHIYUKI MATSUNAGA
Gifu Pharmaceutical University, Gifu, Japan

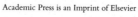

AMSTERDAM • BOSTON • HEIDELBERG • LONDON
• NEW YORK • OXFORD • PARIS • SAN DIEGO • SAN FRANCISCO
• SINGAPORE • SYDNEY • TOKYO

Academic Press is an Imprint of Elsevier

Academic Press is an imprint of Elsevier
125, London Wall, EC2Y 5AS, UK
525 B Street, Suite 1800, San Diego, CA 92101-4495, USA
225 Wyman Street, Waltham, MA 02451, USA
The Boulevard, Langford Lane, Kidlington, Oxford OX5 1GB, UK

Notices
Knowledge and best practice in this field are constantly changing. As new research and experience broaden our understanding, changes in research methods, professional practices, or medical treatment may become necessary.

Practitioners and researchers must always rely on their own experience and knowledge in evaluating and using any information, methods, compounds, or experiments described herein. In using such information or methods they should be mindful of their own safety and the safety of others, including parties for whom they have a professional responsibility.

To the fullest extent of the law, neither the Publisher nor the authors, contributors, or editors, assume any liability for any injury and/or damage to persons or property as a matter of products liability, negligence or otherwise, or from any use or operation of any methods, products, instructions, or ideas contained in the material herein.

British Library Cataloguing-in-Publication Data
A catalogue record for this book is available from the British Library

Library of Congress Cataloging-in-Publication Data
A catalog record for this book is available from the Library of Congress

ISBN: 978-0-12-801918-4

For information on all Academic Press publications
visit our website at http://store.elsevier.com/

Publisher: Janice Audet
Acquisition Editor: Jill Leonard
Editorial Project Manager: Halima Williams
Production Project Manager: Caroline Johnson
Designer: Mark Rogers

Typeset by Thomson Digital

Printed and bound in the United States of America

CONTENTS

PREFACE

This book has required the full support of every member of our biochemistry division, whose cooperation and assistance made its completion possible. The book contains the most recent information on medical biochemistry, except for vitamins. The preparation and writing of this book required not only our intellectual focus but also considerable physical strength and stamina to collect all the information from a wide variety of sources.

Toshiyuki and I are remembering the "Tosa Diary," written in AD 935 under severe conditions by Tsurayuki Kino, who was determined to follow his commitment. In addition, my late son was very involved in the early stages of writing this book. I have composed the following Haiku poem which, together with this book, I consider to be a requiem to his memory:

Fragrant olive smells sweet
when hit by the sun in my yard
facing to the outside.

With the birth of my granddaughter five years ago, my heart has softened and I am now one very happy grandfather.

I would like to thank Mr Brian Paul Lewis for his hard work in helping to review and rewrite this book.

Tsugikazu Komoda, at the clock tower in Kawagoe.

CHAPTER 1

Introduction

Abstract

Biochemistry includes not only biological and chemical materials in the living body, but also homeostasis and cross-communications between biological and chemical materials and components, which may be analyzed by *in vitro* and *in vivo* experiments. Biochemistry investigates the maintenance and preservation of life, addressing the physiological role of biological and chemical materials. The goal of biochemistry is focused on the survival of living organisms.

The processes of life are biochemical. Biological and chemical substances combine in the living body, which not only is an aggregate of biological and chemical materials, but also forms the systems with which an organism can reproduce itself. Thus, the components that manage the metabolism and reproduction of living organisms are more than just the biological materials, but also include structural information: the nucleic acids, lipids, sugars, proteins and enzyme systems of the materials and building blocks of biological and chemical substances that make up life. Precise and delicate chemical systems are found, such as the structure that carries out self-multiplication of the chemical potential used as the metabolic pathway and the driving force of complex systems of energy metabolism and reproducing genetic material. Biochemistry in a broad sense is understood to be an inclusive science for the study of all forms of life, including comprehensive detail of both biology and chemistry. The study of the biomolecules that make up organisms and their environment has become a major field of research in medical science.

Experimentation in Eppendorf tubes (*in vitro*) reveals the biochemical properties of the study and science of biology, and expands our understanding of the structure of all life. Experiments carried out in the living body (*in vivo*) demonstrate the physiological roles in the body. Moreover, since a biomembrane is comprised of membrane proteins and lipids, we see not only the formation of the organelles in the cells, but also previously unknown roles of the cross-communication between biological materials and living cells. This field of study has developed rapidly into the area of biochemistry research.

Biochemistry for Medical Professionals. http://dx.doi.org/10.1016/B978-0-12-801918-4.00001-3

CHAPTER 2

History of Biochemistry

Content

Abstract

The field of study now called biochemistry resulted from the discovery of amylase (diastase) in saliva by Payen in 1833. Basic research on metabolic pathways such as the Krebs (tricarboxylic acid) cycle progressed rapidly. The huge amount of biochemical data derived from experiments using novel biotechnological techniques, e.g. isotope labeling, contributed to the discovery of various metabolic pathways. With the discovery of reverse transcriptase in 1970 came the "Big Bang" of modern biochemistry, leading to the split evolution of molecular biology and cell biology. Two decades later, these subjects reunited to become biochemistry, resulting in the field of regenerative medicine involving induced pluripotent stem cells (iPSCs). Using iPSC techniques, cells originating from another tissue can be formatted and regenerated to target cell membranes or the same organ. The resulting regenerative cells or organ can be transplanted to the damaged organ. Thus, patients can be treated using their own cells.

The dawn of biochemistry came with the discovery of diastase (amylase) as an enzyme by Anselme Payen in 1833. Friedrich Wöhler had reported on the synthesis of urea *in vitro* in 1828, and this result turned the dream that organic matter could be manufactured artificially into a reality. Before then, most scientists believed that organic matter could be synthesized only from and in the living body. Developments in the mid-twentieth century included chromatography, X-ray diffraction, mass spectrometry, high-performance liquid chromatography, isotope labeling, antibody labeling, enzyme labeling, electron microscopic observation of small molecules, molecular dynamics by signal transfer and the discovery of reverse transcriptase (EC 2.7.7.49) as an RNA-dependent DNA polymerase, and these techniques rapidly underwent further development.

In 1970, Howard Martin Temin and David Baltimore separately identified and discovered the RNA-dependent DNA polymerase that catalyzes the reverse transcription reaction of RNA. This discovery was crucial since the enzyme was a catalyst for the reaction of a single strand of RNA as the substrate to complement DNA by reverse transcription. Until this time, scientists had accepted the theory that DNA was biosynthesized by its own duplication and that the genetic code was made only by transcription from DNA to RNA. It became clear with the discovery of this enzyme that the genetic code could also be transmitted to DNA from RNA.

The invention and perfection of technologies using these discoveries led to genetic engineering, which has progressed quickly, resulting in the rapid development of the field of biochemistry. These technological developments have enabled the discovery and analysis of many molecules and metabolic pathways.

Knowledge of biochemistry is used in many fields today from genetics to medicine, dentistry, pharmacy, clinical technology and the science of nursing. The biochemical fields applicable to research can be roughly divided into two: metabolic change and gene expression.

Biochemistry has further developed from molecular biology and molecular genetics, and research has expanded since the early 1980s. Enzymes participate in most metabolic changes in the living body. Since a particular enzyme relates to each stage of a metabolic response, studying the metabolic changes also means studying the enzymes. Regulation of enzymatic function in the signaling pathways triggered by endogenous molecules and chemical substances is also studied in the field of molecular biology.

In 1953, James Watson and Francis Crick discovered the double-helix structure of DNA. This was followed by the discovery and identification of ribosomal protein and gene regulation repressor proteins. These discoveries led to the understanding that both the functions and regulation of the human body are complicated and precise.

Researchers today in the field of biochemistry are undertaking a large amount of research on molecular biology and bioinformatics, to obtain more detailed information on the mechanisms involved in gene expression and interaction. Technology that can create living organisms from stem cells is considered to be within the realms of possibility. The model of induced pluripotent stem cell (iPSC) technology has developed rapidly with few ethical problems as it does not require destruction of the living embryo. It promises to be the epoch-making technology that is capable of regeneration of a part or an entire organ that has been damaged or injured. Regenerative medicine using and applying iPSCs and biochemistry is the next likely step. Regenerative medicine offers incredible opportunities: perhaps in the future we will develop heat-stable functions or the ability to live beneath the surface of the water as we experience global warming!

REFERENCES

Baltimore D. RNA-dependent DNA polymerase in virions of RNA tumour viruses. Nature 1970;226:1209–11.

Payen A, Persoz J-F. Mémoire sur la diastase, les principaux produits de ses réactions et leurs applications aux arts industriels [Memoir on diastase, the principal products of its reactions and their applications to the industrial arts]. Annales de Chimie et de Physique, 2nd Series 1833;53:73–92.

Temin HM, Mizutani S. RNA-dependent DNA polymerase in virions of Rous sarcoma virus. Nature 1970;226:1211–13.

Watson JD, Crick FHC. A structure for deoxyribose nucleic acid. Nature 1953;171:737–8.

Wöhler F. Ueber Kuenstliche Bildung des Harnstoffs. Ann Phys Chem 1828;88:253–6.

CHAPTER 3

Constituents of the Human Body

Contents

Abstract

This chapter describes sugars, glycoproteins, essential amino acids, lipids and enzymes, and their roles as the constituent parts of the human body. Differences between monosaccharides and polysaccharides are described. Glycoproteins consist of sugars attached to proteins; the synthetic pathway of the sugar chain of asparagine-linked sugar moieties is described. The typical roles of 20 amino acids and the taste of amino acids are covered. Concerning fatty acids, the structure of cholesterol and sphingolipids and localization on the cell biomembrane are described. Enzymatic reactions, reaction specificity, the classification of enzymes and isozymes, and the reaction velocity of enzymes are presented. The role played by the cell biomembranes is described.

Biochemistry for Medical Professionals. http://dx.doi.org/10.1016/B978-0-12-801918-4.00003-7

BIOPOLYMERS

Carbohydrates

Monosaccharides and Polysaccharides

Sucrose is the sugar that is used for cooking in the home. It is a disaccharide, consisting of a combination of two monosaccharides: glucose, which is related to the blood sugar level, and fructose, which is abundant in fruit. The sucrose molecule can be decomposed into glucose and fructose by the actions of the enzyme sucrase, which exists naturally in the small intestine.

Polysaccharides are made up of long chains of monosaccharides (Figure 3.1).

Starch is a high molecular weight compound produced by polymerization of α-glucose. There are two types of starch: amylose, with a structure in a straight chain, and amylopectin, with a branched structure. The structure of carbohydrates plays a major role in recognition between biological cells. The structures of protein containing certain sugar chains and lipid-linked sugar chains differ. Different cells have different states in the respective tissues. When recognition does not match at a cell-to-cell level, then certain reactions, such as space structural changes and polarization, cannot take place. Sugar chains have a key role in intercellular recognition. The structure of sugar chains on the cell membrane of cancer cells is different from that

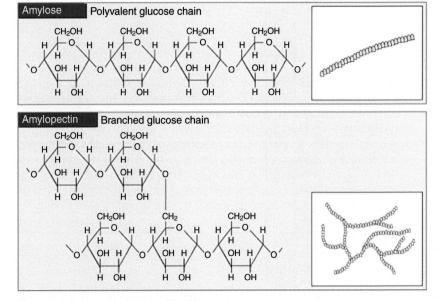

Figure 3.1 *Structure of polysaccharides.*

on normal cells and enables the cancer cells to invade peripheral tissues. Thus, the metastatic actions of cancer are thought to depend on the structure of sugars bound to the membrane. Moreover, certain sugar chains are recognized as cell surface antigens and, for example, bind to receptors for cholera toxin, *Helicobacter pylori* and influenza.

Glycoproteins

A glycoprotein consists of a carbohydrate bound to a protein. Sugar chains combined with glycoproteins can be roughly divided into asparagine (Asn)-linked sugar chains and serine or threonine (Ser/Thr)-linked sugar chains. The Asn-linked sugar chain is a large molecule with a mannose-rich fundamental structure, whereas Ser/Thr-linked chains have a complicated structure and are mainly synthesized in Golgi bodies (Figure 3.2, Table 3.1). During intercellular transport of synthesized proteins and lipids from the endoplasmic reticulum

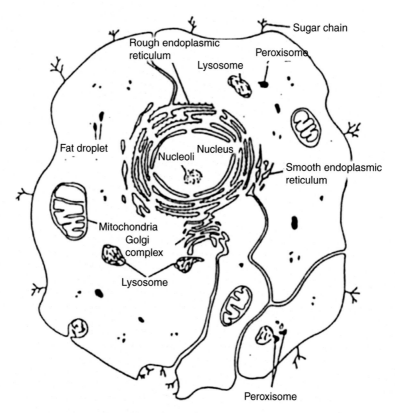

Figure 3.2 *Localization of subcellular organelles.*

Table 3.1 Subcellular fractions and specific markers of organelles

Subcellular fraction	Markers of organelle
Nucleus	DNA
Mitochondria	Succinate dehydrogenase, cytochrome oxidase
Endoplasmic reticulum	Glucose-6-phosphatase, esterase, cytochrome P450
Golgi complex	Thiamine pyrophosphatase, N-acetylglucosamine galactosyltransferase
Lysosome	Protease, acid phosphatase
Peroxisome	Catalase, D-amino acid oxidase
Cell membrane	5′-Nucleotidase, alkaline phosphatase
Cytosol	Lactate dehydrogenase

In general, the specific marker(s) in respective organelles indicate the localized enzyme.

to the Golgi complex, sugars in the organelles are processed by transfer enzymes that recognize the end of sugar chains. Therefore, the structures of sugar chains can differ and are dependent on the expression and localization of transfer enzymes in the Golgi complex. For example, the Asn-linked sugar chains, which initially have a mannose frame, are modified by adding the N-acetylglucosamine, galactose and N-acetylneuramic acid-containing sugar chains to the high mannose type sugar chain in the Golgi compartment. The lipids and proteins modified with the sugar chains are transported to the cell membrane surface, and are involved in key mechanisms between cells.

Amino Acids
Classification and Roles of Amino Acids
Based on their polarity and substituents of the side-chain, the 20 amino acids are generally classified into four groups: non-polar, polar, acidic and basic amino acids.

Non-polar amino acids include nine amino acids: glycine (Gly), alanine (Ala), valine (Val), leucine (Leu), isoleucine (Ile), methionine (Met), phenylalanine (Phe), tryptophan (Trp) and proline (Pro). Of these, Gly, Ala, Val, Leu and Ile are also termed aliphatic amino acids. This is because these five amino acids have structures connected to the hydrocarbon chains. Among the aliphatic amino acids, three (Val, Leu and Ile) with branched hydrocarbon side-chains are also designated branched-chain amino acids (BCAAs) and are known to help recovery from fatigue in the muscles and to provide an energy source during exercise. Gly acts as a neurotransmitter and provides genetic material. Ala is an essential amino acid involved in enhancement of liver function, promotion of recovery from fatigue, and growth hormone that improves activation of macrophages, immune function and reproduction of skin. Sulfur-containing Met is utilized for the biosynthesis of amino acid derivatives such as carnitine, taurine and phosphatidylcholine. Two aromatic amino acids, Phe and Trp, are important precursors to the neurotransmitters dopamine and epinephrine (adrenaline) (from Phe) and 5-hydroxytryptamine (serotonin) (from Trp). Phe is also a precursor to Tyr. Pro is a major constituent of collagen, as well as bones, tendons and ligaments. The common triads observed in the amino acid sequence of collagen are Gly–Pro–X and Gly–X–Pro. Gly is the most abundant amino acid in fibrous protein; the next largest amino acid component in collagen is Pro and its hydroxyl derivative hydroxylproline, which together make up more than 20% of the molecule.

Members of the polar amino acid group are serine (Ser), threonine (Thr), asparagine (Asn), glutamine (Gln), tyrosine (Tyr) and cysteine (Cys). Ser, with a hydroxymethyl group in the side-chain, is enzymatically produced from a glycolytic intermediate, 3-phosphoglycerate, and used in the biosynthesis of nucleic acids, purine and pyrimidine, and Cys. It is also localized in the active center of various enzymes containing trypsin and chymotrypsin and is involved in exerting their enzymatic activity. Thr is an essential amino acid having a hydroxyethyl group and, together with Ser, plays important roles in binding to sugars (O-glycoside linkage) and intracellular signaling (e.g. Ser/Thr kinase). Asn and Gln are produced from their corresponding carboxylic acids, glutamic acid (Glu) and aspartic acid (Asp), respectively. Asn forms not only hydrogen bonds with peptide linkage in proteins but also N-glycoside linkage by binding to sugars. In addition, it participates in forming hydrogen bonding in the secondary structures, the α-helix and β-sheet. Gln is the most abundant amino acid in the body, making up about 60% of the free amino acid pool in skeletal muscles. It is also a source of restorative nutrients in the gastrointestinal tract and intestinal microvilli and is thought to enhance digestive capacity in the intestine and to support the growth and multiplication of immune cells. An aromatic amino acid, Tyr, is produced from Phe and plays a key role in the biosynthesis of thyroid hormones and melanin, as well as the neurotransmitters epinephrine and dopamine. A thiol-containing amino acid, Cys, synthesized from Met, is involved in the formation of disulfide bonds and regulates the activity of tyrosinase, which generates melanin.

Three basic amino acids, lysine (Lys), histidine (His) and arginine (Arg), and two acidic amino acids, Asp and Glu, also help the physiological metabolism of various materials and thereby maintain homeostasis. For example, Lys is involved in glucose metabolism and calcium absorption. His is used as a source for the biosynthesis of histamine and carnosine, which function as an autacoid and a radical scavenger, respectively. Arg supports vessel dilatation and reduction of blood pressure, and improves blood flow via nitric oxide production. Asp and Glu are used as quick-acting energy sources.

Amino Acids in a Meal

Protein is decomposed by proteases and peptidases into small peptides and amino acids, which are absorbed from the gastrointestinal tract and then carried via hematogenous dissemination into the liver. Protein sources include animal proteins contained in meat, fish and eggs, and vegetable proteins in flour. The animal-derived proteins, in general, provide a sufficient balance of nutrients. However, since these protein sources are usually high in fat and have high calorific value, their intake should be balanced with vegetable proteins. Vegetable proteins are degraded to amino acids by the action of multiple enzymes in the fermentation process. Fermented foods from vegetable protein, such as bean paste (miso) and soy sauce (shoyu), which form part of the traditional Japanese diet, do not place a burden on the digestion and are rapidly absorbed. To utilize dietary proteins efficiently, the correct balance of enzymes is needed to digest the proteins. The B vitamins help digestion. With the westernization of eating habits in Japan, few people have amino acid deficiencies. An amino acid deficiency leads to anemia, and a reduction of immune defense mechanisms and muscle mass. Conversely, if humans eat large quantities of meat, the liver and kidneys do not function well. Although ammonia, resulting from amino acid metabolism, can be detoxified into urea in the liver in the urea cycle and excreted as urine, high levels of ammonia confer severe damage such as hepatic encephalopathy in patients with hepatitis and/or nephritis. Such patients may develop other organ disorders, causing alkalosis.

Amino Acids and Taste

A meal may be viewed in terms of its constituent amino acids. The primary role of taste is to determine whether something is edible and nutritious.

The taste buds, which have a diameter of about 0.05 mm, have a similar form to a flower bud and are embedded in fine projections called papillae on the surface of the tongue. When a taste substance binds to a receptor on the cell membrane of a gustatory cell through the villi at its tip, the taste can be sensed only after passing along a nerve fiber via a synapse and sending information to the gustatory area in the cerebrum. Herbivores such as cattle have more taste buds than carnivores, as they need to be able to determine whether a plant is toxic, and the degree of toxicity. For this reason, herbivores need many more taste buds in order to survive.

Table 3.2 Amino acids and taste

Amino acid	Taste
Essential amino acids	
(L-type) Valine (Val)	Bitter
Leucine (Leu)	Bitter
Isoleucine (Ile)	Bitter
Lysine (Lys)	Sweet, bitter
Threonine (Thr)	Sweet
Methionine (Met)	Bitter, sweet
Histidine (His)	Bitter
Tryptophan (Trp)	Bitter
Phenylalanine (Phe)	Bitter
Non-essential amino acids	
(L-type) Alanine (Ala)	Sweet
Arginine (Arg)	Bitter
Glutamine (Gln)	Sweet, salty, bitter
Glutamic acid (Glu)	Acidic, sweet, salty, bitter
Asparagine (Asn)	Bitter, acidic
Aspartic acid (Asp)	Acidic
Serine (Ser)	Sweet, bitter, acidic
Glycine (Gly)	Sweet
Cysteine (Cys)	Sweet, acidic
Proline (Pro)	Sweet, bitter
Tyrosine (Tyr)	None

Taste is divided into categories including sweet, acid, salt, bitter, pungent, umami and sour, and the amino acids have different taste properties (Table 3.2).

The brain can often sense what kind of nutrient is being consumed. It does this almost automatically. For example, a sweet taste may indicate carbohydrate, which is an energy source, or protein, and a salty, sour or bitter taste may indicate poisonous, unripe or rotten food. Newborn infants like the taste of glutamic acid, which occurs in high concentrations in the mother's milk. The taste buds on the baby's face and its breast areas are increased at the third month after birth. Although the embryo takes in nutrients solubilized in the amniotic fluid from the placenta, the full-term fetus is born with taste buds remaining only in the tongue, as in the adult. Since the receptors in gustatory cells cannot combine with large molecules, they cannot taste protein, but can sense and determine various tastes from

Table 3.3 Chemical formulae of some fatty acids

Fatty acid	Chemical formula
Saturated fatty acids	
Lauric	$CH_3(CH_2)_{10}COOH$
Myristic	$CH_3(CH_2)_{12}COOH$
Palmitic	$CH_3(CH_2)_{14}COOH$
Stearic	$CH_3(CH_2)_{16}COOH$
Unsaturated fatty acids	
Palmitoleic	$CH_3(CH_2)_5CH=CH(CH_2)_7COOH$
Oleic	$CH_3(CH_2)_7CH=CH(CH_2)_7COOH$
Arachidonic	$CH_3(CH_2)_4(CH=CHCH_2)_4(CH_2)_2COOH$
Linoleic	$CH_3(CH_2)_4(CH=CHCH_2)_2(CH_2)_6COOH$
Linolenic	$CH_3CH_2(CH=CHCH_2)_3(CH_2)_6COOH$

the amino acids of the form into which the protein is disassembled. There are differences between L-type and D-type amino acids.

Other molecules which contain both amino and acid groups are found in the body that are not included in the 20 amino acids used to synthesize proteins. Examples include γ-aminobutyric acid (GABA), which is a neurotransmitter, and creatine in muscles. The metabolic product of creatine is creatinine, excreted as urinary waste. Threonine (in tea), inosinic acid (in meat, dried bonito, etc.) and guanylic acid (in shiitake mushrooms) provide taste (umami), from ingredients that could be included as natural amino acid derivatives. Ibotenic acid, an amino acid derivative and a neurotoxin, is contained in grisette mushrooms.

Lipids

Fatty acids contain a carboxylic acid with a long hydrocarbon chain. Fatty acids without double bonds in the chain, such as myristic acid and palmitic acid, are designated as saturated fatty acids. In contrast, fatty acids containing one or more double bonds are called unsaturated fatty acids. The chemical formulae of several fatty acids are listed in Table 3.3.

Cholesterol

Cholesterol is a planate polycyclic compound with a hydrophilic hydroxyl residue at the terminal. It is essential for the synthesis of steroid hormones and bile acids in the body. Cholesterols are transported in the bloodstream in particles of various lipoproteins, of which low-density lipoprotein (LDL)-cholesterol is thought to be a major cause of arteriosclerosis.

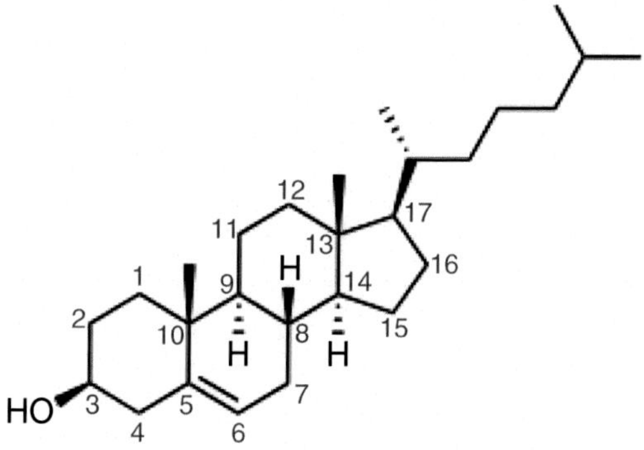

Sphingolipids

Sphingolipids are lipid compounds with an unsaturated hydrocarbon chain and are represented by sphingosine and sphingomyelin. Sphingomyelin exists in the cell membranes of the brain and is a constituent of the myelin sheath, which covers the axial fiber of nerve cells in the form of a membrane. In humans, about 85% of the sphingolipid in the body is sphingomyelin. The sphingomyelin molecule is made up of phosphocholine (red), sphingosine (black) and a fatty acid (blue):

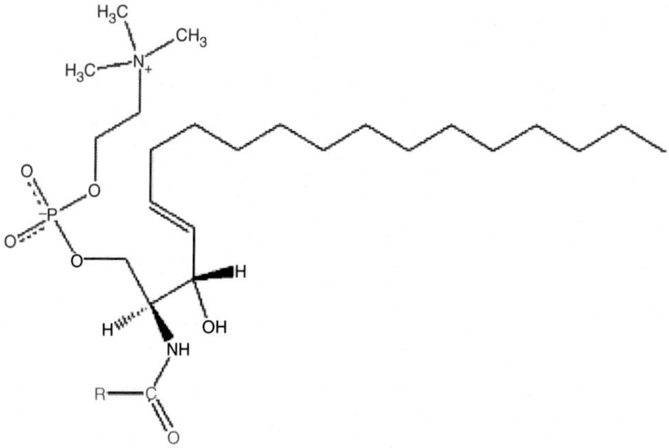

ENZYMES

Enzymes are catalysts that specifically bind to one or molecules, called substrates, and then form an enzyme–substrate complex to yield the product. An enzyme reaction proceeds based on the substrate specificity and kinetic constants. For example, an alkaline phosphatase hydrolyzes organic phosphomonoesters into inorganic phosphate without using any energy.

The model of an enzyme and its substrate was initially compared to the relationship between a lock and a key. In the late twentieth century, the three-dimensional (3D) structure of the enzyme molecule and its "keyhole" was identified by X-ray crystal analysis. There is a cleft, or keyhole, in the enzyme–coenzyme complex. A substrate suitable for the keyhole fits the active center of the enzyme that exists in the inner part of the keyhole and thus produces the metabolite. An enzyme's space structure can be predicted from the base sequence information on the gene that determines its amino acid and space structure. Analysis of the active site is now routinely carried out using X-ray crystal analysis and needs little, if any, primary sequence information on the protein.

It is possible to design proteins or ribozyme enzymes that do not exist in nature. It is also possible to design artificial enzymes using the same technique for substances other than proteins. When the substrate and antibody matching natural enzymes are compounded, it is possible to synthesize the inhibitors/activators of various metabolic enzymes and monoclonal antibodies for metabolic materials. These are excellent tools for the development of medical treatments and molecular biology research [mevalotin, Tamiflu, infliximab, recombinant tissue-nonspecific alkaline phosphatase (TNAP), recombinant intestinal alkaline phosphatase (IAP), etc.].

The isozyme of human lactate dehydrogenase (LDH; EC 1.1.1.27) is a tetramer that consists of four subunits. The subunit composition of LDH varies among different tissues (Table 3.4). The subunit consists of two kinds: cardiac muscle type (B) and skeletal muscle type (A). When any of these four subunits are combined, LDH is constituted. Therefore, although five forms of the LDH isozyme exist, these relate to the isozyme that controls the same biochemical reaction with the same substrate. If this model is applied, an organ disorder can be identified from electrophoretic data analysis of the isozyme type of LDH. Two or more enzymes that control a series of metabolic processes form a cluster. The enzyme becomes a conjugated enzyme

Table 3.4 Human lactate dehydrogenase and isozyme types

Type	Subunit composition	Histological distribution
LD1	M_4	Heart
LD2	HM_3	Skeletal muscle, diaphragm, kidney
LD3	H_2M_2	Leukocytes
LD4	H_3M	
LD5	H_4	Liver

in many cases. The representative conjugated enzymes in a fatty acid synthesis system are acyl carrier protein (ACP) complex, which conjugates with six enzymes, [ACP] S-acetyl transferase, maronyltransferase, 3-*oxo*-acyl ACP synthase, 3-*oxo*-acyl ACP reductase, crotonyl ACP hydratase and enoyl ACP reductase. Most enzymes in the fatty acid synthesis system are conjugated; the only independent enzyme is acetyl coenzyme A carboxylase.

Enzyme Reaction Speed

In the first phase of an enzyme reaction, if the substrate concentration [S] increases, the reaction velocity of the enzyme will increase. If a substrate concentration is changed, the reaction velocity will create a hyperbola that results in the saturation maximum velocity or rate (V_{max}) (Figure 3.3). In the case of a metallic catalyst, it is not dependent on the catalyst concentration and the reaction initial velocity [v] is determined by the linear expression of substrate concentration [S].

In general, the enzyme molecule is considered to be huge compared with many substrates and does not have more than one active center. Therefore, the reaction between the substrate and enzyme is slower compared to the reaction with a metallic catalyst. If the substrate concentration increases in an enzyme reaction, saturation will take place (V_{max}). Thus, in the process of a substrate conjugating to an enzyme, the substrate cooperates with an enzyme in the enzyme reaction. This is considered to be a rate-limiting process. It could be a regulated reaction velocity of an enzyme reaction.

In 1913, Michaelis and Menten measured the hydrolysis reaction of sucrose by sucrase. From their results, they presented a model of the

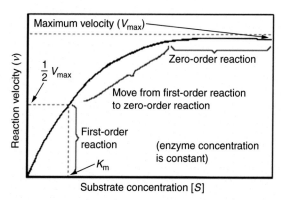

Figure 3.3 *Enzyme reaction.* V_{max}: maximum velocity; K_m: Michaelis–Menten constant.

enzyme–substrate complex from the lock and key model, and formulated the enzyme reaction as follows:

$$E + S(ES) \rightarrow E + P \tag{3.1}$$

An enzyme reaction is divided into two steps. In the first step, the enzyme and its substrate are connected temporarily and form an enzyme–substrate complex. In the second step, the enzyme–substrate complex divides into an enzyme and a product.

Inhibition Style and Enzyme Reaction Speed

The reaction velocity of an enzyme is affected by the existence of molecules resembling the substrate and any allosteric effects. The inhibitory mechanism can be investigated and determined according to the type of inhibition. Thus, the enzyme reaction can be predicted from the type of regulatory enzyme present.

By studying the regulatory controls of enzyme actions, new medicines can be developed to treat pathological conditions.

An inhibition style may be classified as follows: L-Phe as a non-competitive inhibitor for human IAP, L-HomoArg as an irreversible inhibitor (non-competitive inhibition) for human TNAP, and L-Leu as a reversible inhibitor (competitive inhibition) for human IAP or human PAP. The reciprocal plots of both sides of the Michaelis–Menten formula (Equations 3.2 and 3.3) follow (Figure 3.4):

$$v = \frac{V \cdot [S]}{K_m + [S]} \tag{3.2}$$

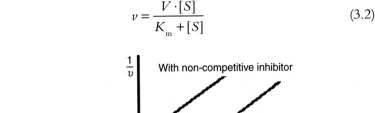
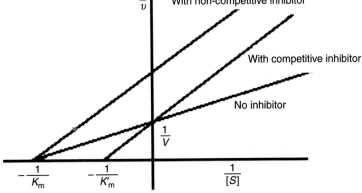

Figure 3.4 *Lineweaver–Burk plot: relation between a competitive inhibitor and a non-competitive inhibitor.*

Table 3.5 Activation energy

Reaction name	Catalyst/enzyme	Energy (aq.) (kcal/mol)
Hydrolysis of casein	Trypsin	20
	Chymotrypsin	12
Hydrolysis of lipid	Lipase (pancreas)	4.2

aq.: aqueous.

$$\frac{1}{v} = \frac{K_m}{V} \cdot \frac{1}{[S]} + \frac{1}{V} \tag{3.3}$$

A chemical reaction generally advances in the direction in which chemical potential is decreased, i.e. the direction that consumes energy. Reaction velocity is concerned with whether the activation energy of a reaction is high (Table 3.5). There is uncertainty about the detailed molecular basis of enzyme reactions, such as the mechanisms underlying substrate specificity, and the lowering of activation energy by adding inorganic and acid–base catalysts.

Recent recombinant techniques such as mutagenesis have revealed that an enzyme–substrate complex may be formed due to entropy trap. The enzyme reaction is limited and determined by the induced fit in the active center, and the reaction proceeds to output. A good example is chymotrypsin, a type of serine protease, named after the serine playing a role in the mechanism. His57 in the protease interacts with negatively charged Asp102 to transfer the proton. Next, His57 acts as a base and acquires a proton from an active center residue Ser195 to enable it to attack a substrate. The enzyme reaction is then terminated by sequential proton recovery (substrate → His57, and Asp102 → His57).

An enzyme reaction converts a substrate into a product with progress from a transition state to a ground state through the intermediate states, which are dependent on the molecular structure, such as bond distances and atomic angles. In other words, there is an induced fit resulting from the chemical reaction that occurs in the enzyme–substrate response. Since an enzyme reaction is equivalent to the higher potential state, the enzyme reaction can reach the intermediate state with little energy. If the primary role of an enzymatic protein is to make a transition state, the transition state is thought to be achieved by combining the enzyme and substrate.

Like enzymes, an antibody is capable of distinguishing its counterpart among the biological material. In 1986, Tramontano and colleagues discovered an antibody with an enzymatic action, which they named an abzyme.

Classification of Enzymes by Mechanism

Enzymes can be classified in different ways, including by the location of the enzyme, the kind of substrate (difference between substrate specificity and reaction singularity) and the enzyme reaction. Classification by the last method is connected with the nomenclature of enzymes.

All reactions in the body take place as a result of the actions of enzymes. Therefore, if an enzyme is localized where a metabolic response is involved, the enzyme exists universally and is classified into membrane enzymes bound with the biomembrane (cell membranes and the membranes of cell organelles) and soluble enzymes that exist in the cytoplasm or luminal fluid. Some soluble enzymes are secreted out of cells. Differences between enzymes are based on the difference in location on the structures of the hydrophobic and hydrophilic side-chains, based mainly on the amino acid sequence.

Although enzymes, as well as other proteins, are biosynthesized by intracellular ribosomes, the evolution of enzymes may be reflected in the structure because the motif is dependent on a particular gene. Hereditarily, similar enzymes have a similar motif and form an enzyme group.

Membrane Enzymes

Many enzymes in the biomembrane participate in the conservation of energy or transportation of substances making up important enzyme groups (5'-nucleotidase, γ-glutamyltransferase, etc.). The membrane enzymes can be divided into three kinds according to the spatial relationship of the biomembrane and the enzyme. The first type, integral membrane enzymes, is buried in the biomembrane (receptor proteins, etc.); the second penetrates the biomembrane (channel transporter, etc.); and in the third type, peripheral membrane enzymes, the enzyme is attached to the biomembrane (alkaline phosphatase, γ-glutamyltransferase, etc.). Since the biomembrane has a hydrophilic surface and an inner hydrophobic fraction (lipid bilayer membrane), the membrane anchoring moiety (hydrophobic peptide chain) of the enzyme links with the membrane by strong hydrophobicity. Regarding the affinity to the membrane, the lipid moiety also has very strong affinity to the membrane as a glycolipid moiety (GPI anchor).

Soluble Enzymes

Cytosolic enzymes exist in their solubilized form in the liquid phase in cells or luminal fluid in the bloodstream. Many of these soluble enzymes are concerned with metabolism in the cytoplasm. Many hydrophilic amino acids are positioned on the outside of soluble enzymes and hydrophobic amino acids are found on the inside of the structure.

Although enzymes are produced within certain cells, some enzymes, called secretory enzymes, are secreted out to the cytosol fraction after formation. For example, digestive enzymes digest substances to facilitate their absorption into the cells. Many secretory enzymes are excreted in soluble forms. If a certain stimulus is applied to the body, secretory enzymes are released through exocytosis. Research in structural biology has identified the 3D structure of soluble enzymes by crystallization techniques.

Classification of Enzymes in Catalysis

Enzymes can be classified systematically according to the difference between reaction and substrate specificity, and the mechanism of action. The enzyme code (EC) shows such a classification. Notation according to EC number, i.e. EC X.X.X.X, is shown as follows. The first EC number classifies the enzyme reaction mechanism into six groups, namely oxidation–reduction, transition, hydrolysis, dissociation, isomerization and synthesis (creating new chemical bonds with the initial assistance of ATP). Examples of enzymes classified by EC number are:

- EC 1.X.X.X-reductase
- EC 2.X.X.X-transferase
- EC 3.X.X.X-hydrolase
- EC 4.X.X.X-lyase
- EC 5.X.X.X-isomerase
- EC 6.X.X.X-ligase.

Although standards of classification differ in each group, they are subdivided by the difference between enzyme reaction and substrate specificity. These numbers are assigned to the entire enzyme, and over 3000 enzyme reactions have been assigned an EC number. Moreover, enzymes have a variety of activities; for example, ATPase catalyzes the hydrolysis of both proteins and ATP. Sometimes, the substrate that the enzyme metabolizes is omitted from the systemic naming, based on the same rules as a systemic name. For example, the systemic name of EC 1.1.1.1 is alcohol dehydrogenase. There are also many enzymes named in accordance with the naming convention, such as DNA polymerase.

Although an enzyme generally consists of protein, a few enzymes contain non-protein components such as nucleic acid. The ribozyme discovered by Thomas Cech and others in 1986 is a catalyst made of RNA, which acts on itself and cleaves RNA.

Some enzymes require other molecules to function and do not become active unless combined with cofactors (coenzyme, metal, etc.). An apoenzyme, a protein portion without a cofactor, does not have enzymatic activity, whereas a holoenzyme, a protein combined with a cofactor, has such activity.

The organic compound of the non-protein which assists an enzyme reaction in the active center is called a coenzyme. Since coenzymes are essential elements in the active form of the enzyme, they belong to a prosthetic group. Although they differ from typical prosthetic groups, they can easily separate from the enzyme and be consumed during an enzyme reaction with a substrate like NADH. For example, since the cytochrome P450 (CYP) enzyme is bound covalently with the heme iron, heme does not separate from the CYP enzyme. Therefore, this heme moiety is not called a coenzyme. Although lipoic acid is bound covalently to the enzyme, lipoic acid can be separated from the enzyme moiety, so lipoic acid is called a coenzyme. Therefore, the criteria defining coenzymes and prosthetic groups are not strict.

An enzyme may be comprised of two or more protein chains (peptide chain). When it consists of two or more peptide chains, each peptide chain is called a subunit.

BIOMEMBRANES

The main constituents of a biomembrane are proteins and lipids. Their ratio varies in the body and the organs. The constituted lipid of biomembranes includes an important steroid and phosphatide located in the core. In an animal cell, cholesterol is the major biomembrane component, whereas the membrane in plant cells has a higher level of sitosterol. Glycolipids are also found in the membrane, with the sugar chains bound to membranous protein or lipid facing the extracellular surface of the cells. In glycerolipids, phosphatidylcholine, phosphatidylethanolamine, phosphatidylserine, etc., are major components of membranous lipids. As a kind of phosphatide, glycerophospholipids have a glycerol frame and sphingophospholipids have

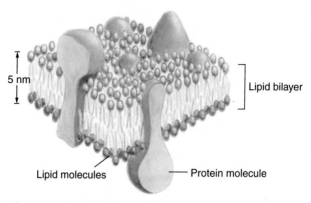

5 nm

Lipid bilayer

Lipid molecules

Protein molecule

Figure 3.5 *Bilayer structure of a biomembrane.*

a sphingosine frame. Sphingomyelin is a major component of sphingo-phospholipids. If a biomembrane were merely a physical wall separating the inside and outside of the cell, it would not need to consist of various lipids. Biomembranes consist of over 10 kinds of lipid that perform different roles, including transmission of signals by metabolic enzymes, modulation of receptors, selective transparency by transporters, active transport, facilitated diffusion, immune response and respiratory chain complexes.

Figure 3.5 shows the biomembrane model advocated by Singer and Nicolson (1972). In this biomembrane, protein is anchored in the liquid mosaic consisting of a lipid bilayer, and membrane protein anchors from the outside to the inside. In addition, the membrane protein is moved by lipid fluidity. Both the lipid orientation and protein anchoring in biomembrane can be supported by this model. The channel part of the membrane protein is arranged within membrane lipids, and the hydrophilic portion is arranged on the outside of the membrane. The proteins are not completely fixed, but can undergo conformational changes in the membrane, and this can be observed using technology such as single-molecule observation. The lipid bilayer is a membrane-like structure that includes polar lipids, especially phosphatide, to form two layers. It is one of the biomembrane models of Danielli and Davson (1935), who proposed that "a cell membrane consists of a double layer of lipid."

The polar lipid in a lipid bilayer structure is amphiphilic, having both hydrophilic and hydrophobic properties. A hydrophilic group gives the soluble molecule polarity, electrostatic combination or a hydrogen bond, and faces the plasma side. Hydrophobic groups condense into a group so that water can be avoided.

An amphiphilic molecule forms a lipid bilayer as one of the most stable structures under liquid. The side end of one lipid double membrane and the side end of another are united, and an unreserved hydrophobic group is separated from a soluble molecule. A spherical short segment where a side end does not exist in a bilayer is formed. This spherical lipid double membrane is called a lipid bilayer vesicle, and the inside is filled with the soluble molecule.

A pure planar unit is removed when a side end is isolated from a soluble molecule by endocytosis. The structure of biomembranes is very complicated as the anchoring protein links with various transmitters, molecular motors and adhesive proteins.

REFERENCES

Cech TR. A model for the RNA-catalyzed replication of RNA. Proc Natl Acad Sci USA 1986;83:4360–3.

Danielli JF, Davson H. A contribution to the theory of permeability of thin films. J Cell Comp Physiol 1935;5:495–508.

Singer SJ, Nicolson GL. The fluid mosaic model of the structure of cell membranes. Science 1972;175:720–31.

Tramontano A, Janda KD, Lerner RA. Chemical reactivity at an antibody binding site elicited by mechanistic design of a synthetic antigen. Proc Natl Acad Sci USA 1986;83:8736–40.

CHAPTER 4

Metabolic Pathways in the Human Body

Contents

Biochemistry for Medical Professionals. http://dx.doi.org/10.1016/B978-0-12-801918-4.00004-9

Abstract

Major metabolic pathways for several biological materials are described, including carbohydrate and energy metabolism by electron transfer systems, lipids, lipoproteins, amino acids, nucleic acid and protein biosynthesis. Metabolic syndrome is caused by disruption of metabolic pathways or their regulation. Disorders in anaerobic sugar metabolism and glycogen metabolism can cause diabetes mellitus. Alzheimer and Parkinson disease are examples of disorders of electron transfer systems. Concerning the nature of sugar chains, certain blood group substances are useful as tumor markers. Dysfunction of the glycosylphosphatidylinositol moiety binding to asparagine-linked sugar chains causes paroxysmal nocturnal hemoglobinuria. Lipid metabolism is an important indicator of lipoprotein and fatty acids; both cholesterol and lipid metabolism have relevance to disease. Disorders of amino acid metabolism and nucleic acid metabolism and the resulting diseases are illustrated. An overview of protein synthesis is provided.

SUGAR METABOLISM

The mechanism of absorption of sugar in the small intestine is shown in Figure 4.1.

Glycolysis

Preparatory Stage

The five steps in the first half of glycolysis are called the preparatory or energy investment phase. In the preparatory phase, two molecules of adenosine triphosphate (ATP) are supplied, and the phosphorylation of glucose and conversion to glyceraldehyde–3–phosphate (G3P) take place.

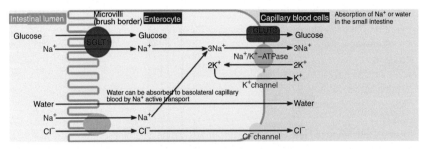

Figure 4.1 *Mechanism of absorption of sugar in the small intestine.* Glucose is absorbed by type 1 glucose transporter (GLUT1), which is located in microvilli in the proximal enterocytes. Then, elevated glucose in the enterocytes can be taken into the bloodstream by type 2 glucose transporter (GLUT2). SGLT1: type 1 sodium-dependent glucose transporter; NCC: Na^+/Cl^- cotransporter.

Stage 1: Phosphorylation of Glucose

Glucose is phosphorylated with the enzyme hexokinase in the first step of the preparatory phase. ATP is a donor of a phosphate group; the carbon 6 (C6) position of glucose transfers to a γ-phosphate residue, and glucose-6-phosphate (G6P) is generated. This reaction requires magnesium ions (Mg^{2+}). Hexokinase has several forms (isozymes). Although the catalyst advances the same reaction, the isozymes are coded by different genes. Mammals have four isozymes of hexakinase. These isozymes have different affinities (Michaelis constant or K_m value) for glucose. The K_m value of hexokinase I, II and III is 10^{-6} M. The K_m of hexokinase IV (also called glucokinase) is usually 10^{-2} M, depending on the role of glucokinase. A large concentration of glucokinase exists in hepatic cells. Since the glucose concentration in blood is lower than the K_m, glucokinase does not usually act fully against glucose. In this case, other isozymes of hexokinase catalyze the reaction. However, when the glucose concentration increases, glucokinase will begin to act. Since glucokinase is hardly saturated, even if the glucose concentration in hepatic cells increases markedly, glucose can move rapidly to the glycolytic pathway or glycogen synthesis shunt. The intracellular glucose concentration is always low, and extracellular glucose inhibits the flow of glucose to the outside of the cell and promotes membrane transport into the cell. Insulin can promote the action of hexokinase. In contrast, glucagon can repress the action of hexokinase, resulting in the acceleration of glucose-6-phosphatase activity. Then, the gluconeogenesis reaction will be promoted *in vivo*.

Stage 2: Isomerization of Glucose-6-Phosphate

Glucose-6-phosphate (G6P) is changed into fructose-6-phosphate (F6P) by phosphoglucoisomerase (phosphoglucose isomerase) in the second step. This reaction also requires Mg^{2+}. F6P can enter the glycolytic pathway from the next point. This reaction has a low free energy charge. It can progress in both directions, but since F6P is rapidly consumed in the following step, it is unlikely to undergo the reverse reaction. The α-anomer of G6P, α-D-glucopyranose-6-phosphate, combines with the enzyme preferentially, the ring opens and it converts an aldose to a ketose. This isomerization reaction is important for the glycolytic pathway after this step.

Stage 3: Phosphorylation of Fructose-6-Phosphate

In the third step, phosphofructokinase-1 (PFK-1) transfers the phosphate residue of ATP to the C1 hydroxyl residue of F6P, generating fructose-1,6-bisphosphate (F1,6BP). This reaction requires Mg^{2+} and fructose-2,6-bisphosphate. After ingestion of a meal, fructose-2,6-bisphosphate is elevated, resulting in the above-mentioned reaction advancing a speedy glycolysis reaction. Before a meal, the concentration of fructose-2,6-bisphosphate is decreased, resulting in upregulation of F1,6BP and facilitation of gluconeogenesis. Although G6P and F6P are also metabolized in pathways other than the glycolytic system, F1,6BP is metabolized only by the glycolytic pathway. The reaction in which PFK-1 acts as a catalyst is irreversible, and PFK-1 does not use the reverse reaction in the case of glyconeogenesis. Therefore, two enzyme reactions modulated by fructose-2,6-bisphosphate are a key point of the glycolytic and/or glyconeogenesis system.

Stage 4: Cleavage

The first three reactions of glycolysis are the preparative stages for cleaving F1,6BP and making two triose phosphates. In this step, F1,6BP is cleaved by F1,6BP aldolase (or just aldolase) into G3P and dihydroxyacetone phosphate (DHAP). This reaction, catalyzed by aldolase, results in a positive large standard free energy change in the direction in which F1,6BP is cleaved, within the cells. The reaction is mostly in equilibrium and is not a control point of glycolysis. When the intracellular metabolite concentration is low, the free energy change is low and a reverse reaction occurs easily. There are two types of aldolase: type I aldolase exists in animals and plants, and type II aldolase in fungi and bacteria. The two types differ in the cleavage mechanism of hexose.

Stage 5: Isomerization of Triose Phosphate

Of the two molecules described in the preceding paragraph, only G3P serves as a substrate continuing in the next steps of the reaction. However, DHAP is promptly changed into G3P by a reversible reaction of triose phosphate isomerase. Since triose phosphate isomerase acts as the catalyst of a reaction on a solid unique target, only the D-isomer is generated. The carbon 1 (C1) position of the G3P is derived from the C3 of DHAP, and the C2 and C3 positions of the G3P originate from the C2 and C1 positions of glucose. Of the C1, C2 and C3 positions of the glyceraldehyde-3-phosphate generated in step 4, the above C3 is the C4, C5 and C6 positions of glucose. However, the distinction does not separate chemically the carbon of each position of the two G3Ps. Two molecules of G3P are generated by this reaction from a hexose molecule, and the preparatory phase of glycolysis ends.

Payoff Phase

The five steps in the second half of glycolysis are called the payoff phase or energy phase. In the payoff phase, two molecules of G3P are changed into pyruvate, four molecules of adenosine diphosphate (ADP) per molecule of glucose are changed into four ATP, and a part of the free energy of glucose is saved. Since two molecules of ATP were consumed in the preparatory phase, the net gain of ATP through glycolysis will be two molecules. Moreover, two molecules of nicotinamide adenine dinucleotide (NADH) per molecule of glucose are generated.

Stage 6: Oxidation of Glyceraldehyde-3-Phosphate

In the step at the beginning of the payoff phase, G3P is changed into 1,3-bisphosphoglycerate in a reaction catalyzed by glyceraldehyde-3-phosphate dehydrogenase. Dehydration of the aldehyde residue of glyceraldehyde-3-phosphoric acid takes place, and one molecule of the oxidized form of nicotinamide adenine dinucleotide (NAD^+) is changed into the reduced form (NADH). Oxidation of the aldehyde group in G3P decreases the standard free energy and much of the decreased energy is saved in an acyl phosphate. Acyl phosphate is a substance with an ester bond of a carboxylic acid with an acyl residue (R—CO—) and phosphate, and the energy release at the time of hydrolysis is very rapid. This energy is required to generate ATP from ADP in the following stage. When heavy metals such as mercury (Hg^{2+}) react with the cysteine of the active center of an enzyme, an enzyme reaction is stopped irreversibly. Hg^{2+} poisoning induces renal failure. Since the quantity of NAD^+ in the cell is limited, if NADH oxidizes again and NAD^+ is not supplied continually, a reaction will stop. Alcoholic and lactate fermentation are examples of reactions by which NADH can be regenerated by reoxidation.

Stage 7: Transition of the Phosphate Residue from 1,3-Bisphosphoglycerate to Adenosine Diphosphate

The second step of the payoff phase is a phosphoglycerate kinase (PGK)-based reaction. The phosphate residue, having higher energy, transfers from 1,3-bisphosphoglycerate to ADP in a reaction catalyzed by PGK with generation of ATP and 3-phosphoglycerate. This is substrate-level phosphorylation in glycolysis. The reaction requires Mg^{2+}. Oxidation of aldehyde in stage 6 and generation of ATP in stage 7 are conjugated, and constitute the energy coupling reaction which uses 1,3-*bis*-phosphoglyceric acid as an intermediate form. The formation reaction of the first acyl phosphate action is an endergonic reaction, and generation of the following ATP is a strongly exergonic reaction. If the two reactions are summed, the energy can be expressed with the following formula:

$$\text{Glyceraldehyde-3-phosphate} + ADP + Pi + NAD +$$
$$\text{3-Phosphoglycerate} + ATP + NADH + H^+$$

where P_i is inorganic phosphate. Overall, this is an exergonic reaction.

Stage 8: Intramolecular Rearrangement of Phosphate Residue

The reaction in this step is an intramolecular rearrangement, not by isomerization but by mutase. The phosphate residue of glycerate is transferred reversibly by the catalyst phosphoglycerate mutase (PGM), and is changed into 2-phosphoglycerate. This reaction requires Mg^{2+}. This reaction consists of two steps. The reaction mechanisms differ between animals and plants. In animals, phosphate bound to the His residue of the active center in the enzyme transfers to the hydroxyl residue of the C2 position of 3-phosphoglycerate to yield 2,3-bisphosphoglycerate (2,3-BPG or just BPG) as the reaction intermediate. Next, while the transferred phosphate residue is returned back to the His residue from C3 of 2,3-BPG, reproduction of the phosphorylated enzyme and formation of the reaction product 2- phosphoglycerate are carried out. Because 2,3-BPG is required to phosphorylate PGM, some 2,3-BPG must always be stored in cell to initiate the catalyzed reaction. By contrast to the case in animals, the intermediate 2,3-BPG is not produced in plants. Briefly, 3-phosphoglycerate binds with the enzyme first and a phosphate residue transfers to the active center. 2-Phosphoglycerate is then generated by returning this phosphate residue to C2.

Stage 9: Dehydration of 2-Phosphoglycerate

In the fourth step of the payoff phase, a compound with high-energy phosphate transition potential is generated again. The reaction is catalyzed by enolase, which reversibly uses water (H_2O) from the C2 and C3 in 2-phosphoglycerate, resulting in conversion to phosphoenolpyruvate (PEP). Since enol in pyruvate is unstable, the phosphate transition potential is very high. In fact, the standard free energy required for hydrolysis of the phosphate group in 3-phosphoglycerate is $\Delta G = -17.6$ kJ/mol and that in PEP is $\Delta G = -61.9$ kJ/mol. Although the total amount of calories is almost the same in both compounds, redistribution of energy takes place in the dehydration reaction. To advance this reaction, the participation of two Mg^{2+} is needed. One Mg^{2+} binds to the carboxyl residue of a substrate as the "conformational" ion, while another Mg^{2+} binds to a carboxyl residue and a phosphate residue as the "catalyst" ion. These two ions negate oxidation, and the lysine residue of the active center of enolase draws out the hydrogen atom at the C2 position, and the hydrogen atom of the glutamate residue of enolase and the hydroxyl group from the C3 position from H_2O.

Stage 10: Transition of Phosphate Residue from Phosphoenolpyruvate

The enolic form of pyruvate is generated by the action of the transition pyruvate kinase in the cells, where it is quickly converted into the keto form, an isomer of PEP.

The last step in the glycolytic pathway is transition of the phosphate residue from PEP by the catalyst pyruvate kinase to ADP, and the resulting pyruvate and ATP are generated in this phase. This reaction requires either potassium (K^+) and Mg^{2+} or manganese (Mn^{2+}) ions. The phosphorylation reaction depends on the level of the second substrate. This reaction is irreversible, depending on the intracellular environment, and is an important regulatory point. Approximately half (30.5 kJ/mol) of the energy released by hydrolysis of the anhydrous phosphate of PEP is saved as ATP.

Location of Glycolysis

The reaction of glycolysis occurs in the cytoplasmic matrix in cells. This is probably because glycolysis is the most primitive metabolic system, existing before intracellular organelles evolved. In a eukaryotic organism, metabolites obtained by glycolysis are carried into the mitochondria, where they supply the reactions of the tricarboxylic acid (TCA) cycle (also known as the citric acid or Krebs cycle) and electron transfer system to produce energy under aerobic conditions.

Role of Glycolysis

Glycolysis is the most fundamental system for sugar metabolism in the body. It contributes to the production of the energy currency ATP, as well as NADH, which is used to create ATP in the electron transfer system. Lactate and ethanol fermentations are the shunts for maintaining glycolysis, and reproduce NAD^+, essential for the processes of glycolysis, using the NADH formed. The pyruvate obtained by glycolysis is used in the TCA cycle only after pyruvate has bound with coenzyme A (CoA) to obtain acetyl-CoA. Glycolysis has two roles. The first role is the generation of ATP. In addition to ATP production in glycolysis, metabolism in the TCA cycle and oxidative phosphorylation of acetyl-CoA supply much more ATP. In fact,

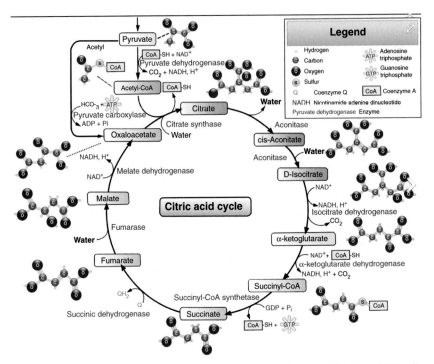

Figure 4.2 *Tricarboxylic acid (citric acid or Krebs) cycle. Source: Modified from Wikipedia.*

only two moles of ATP per mole of glucose are produced under anaerobic conditions, whereas about 38 moles of ATP can be produced under aerobic conditions. The second role is the formation of intermediate metabolites, used as precursors for many biosynthetic pathways. For example, acetyl–CoA is a precursor for fatty acid synthesis. Acetyl–CoA is synthesized from proteins, sugars and lipids, as well as from pyruvate by joining the acetyl residue and CoA together. As illustrated in Figure 4.2, acetyl–CoA produced in this way goes into the TCA cycle, which consists of nine steps in the matrix of mitochondria. The enzyme complex of the mitochondrial inner membrane performs several reaction steps. The end-product of glycolysis, pyruvate, is changed into acetyl–CoA by a combination of decarboxylation and CoA. Acetyl–CoA is also obtained from β-oxidation of fatty acids and metabolism of amino acids. Under anaerobic conditions, pyruvate is converted into lactate by lactate dehydrogenase. In patients with acidosis, such as in hyperventilation or diabetes, lactate is converted to pyruvate. Therefore, the reaction of lactate dehydrogenase is reversible and depends on the levels of oxygen *in vivo*.

Metabolism of the Tricarboxylic Acid Cycle and Electron Transfer System

The TCA cycle oxidizes the acetyl group of acetyl-CoA, conferring two molecules of carbon dioxide (CO_2), and generates two coenzymes [NAD^+ and flavine adenine dinucleotide (FAD^+)] to their respective reduced forms ($NADH^+$ and $FADH_2$). In addition to the degradation of sugars, it contributes to many other pathways, such as amino acid metabolism, the urea cycle and glyconeogenesis. Unlike anaerobic glycolysis (Embden–Meyerhof–Parnas or EMP shunt), it should be noticed that ATP is not produced directly in the TCA cycle. Except for conversion of GTP into ATP in the cycle, the reduced coenzymes ($NADH$ and $FADH_2$) generated are used for synthesis of ATP by oxidative phosphorylation following the electron transfer system. These sugar metabolism pathways take place in the cytosol and mitochondria of hepatic parenchymal cells. Since glucose serves as a substrate of anaerobic glycolysis, skeletal muscle can perform high-level activity even when aerobic oxidation cannot be used, so that an organism can function even in an oxygen-poor environment.

Regulation of Glucose in the Bloodstream

Glucose in the blood can be obtained from three sources: meals, glycogen degradation and glyconeogenesis. During fasting, glycogen is first degraded in the liver and the kidney to release glucose into the bloodstream. Distinct from that, in the liver, the muscle glycogen is used to supply an energy source for muscles. However, the amount of glycogen stored in the liver is limited and the glycogen-derived glucose is rapidly depleted.

When the fasting continues for a long time, glucose is biosynthesized, by glyconeogenesis, from glycerol elicited by metabolizing various substances, i.e. glycogenic amino acids (by a cycle of protein breakdown) and lipids. Although the glyconeogenesis reaction is somewhat slower than the corresponding decrease in blood sugar level, glucose is synthesized and supplied continuously.

Regulation of the flow of the whole metabolic pathway is important to maintain a steady supply of the chemicals required for the pathway. This is achieved by regulation of only one reaction in which catalysis takes place, or two reactions with the "regulatory enzyme" in metabolism. The activity of some regulatory enzymes is controlled by hormones. The metabolism and the hormonal action of monosaccharides, proteins and lipids have been described. Glucose metabolism is normally regulated by the action of insulin, glucagon, glucocorticoid, growth hormone and epinephrine (adrenaline).

Levels of stress hormones such as epinephrine and cortisol are increased under stressful conditions, and those of anti-insulin hormones are involved in the pathogenesis of stress hyperglycemia. The completely different intermediary metabolism during an absorption period and a fasting period is mainly adjusted by hormones. Insulin works during the absorption period, and epinephrine, glucagon, growth hormone, thyroid hormone, adrenocortical hormone, etc., work during the fasting period. In patients with type 1 diabetes mellitus, the blood level of insulin is decreased owing to congenital abnormality of the cells or enzymes involved in its secretion. In contrast, a reduction in sensitivity to the hormone insulin is an abnormal response (dysfunction of the receptor) and is defined as type 2 diabetes; the blood sugar level in these patients appears to improve with the administration of antidiabetic medicines including acarbose.

An electron transfer system is a reaction system in the final stage of two or more metabolic systems during aerobic respiration. It is also called a hydrogen transfer system, respiratory chain, etc. Its purpose is to make the inside and the outside of a biomembrane produce a gradient of proton concentration, and then ATP enzymes for synthesis generate ATP using this proton concentration slope (Figure 4.3). Cytochrome c is located at the membrane surface, although most of the proteins that constitute a respiratory chain or coenzyme groups are embedded in the biomembrane. The proton carried as NADH produces complex I. Mutations can cause a deficiency of the mitochondrial electron transfer system known as MELAS, a syndrome that occurs in 3000 newborn infants every year. Electrons carried as $FADH_2$ are also passed to a ubiquinone (coenzyme Q) from complex II.

Hydrogen ions ($2H^+$) in the reduced type of ubiquinone are moved to the intermembrane space by interacting with complex III and, simultaneously, an electron is passed to complex III. The electron passed to complex III is sent to complex IV through the cytochrome c of the superficial mitochondrial membrane. Complex IV oxidizes the reduced type of cytochrome c, and the electron is passed to an oxygen (O_2) molecule. If O_2 combines with two H^+ in a matrix, one molecule of water will be synthesized. In total, 10 H^+ are carried to the intermembrane space through the crevice between complexes by complexes I, III and IV. The gradient of H^+ concentration formed in the proton transfer raises the membrane potential. Aerobic respiration requires O_2 because O_2 is required in the last reaction of an electron transfer system.

The deadly poison potassium cyanide inhibits complex IV in the electron transfer system. A deficiency of complex IV leads to Parkinson disease

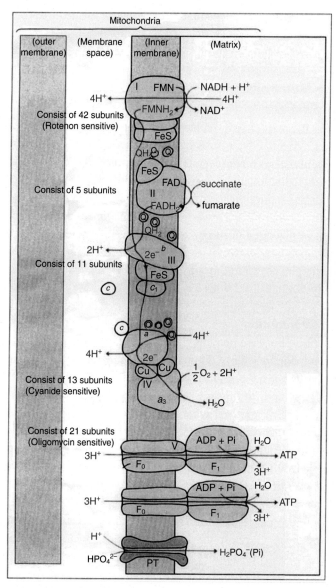

Figure 4.3 *Electron transfer system and adenosine triphosphate (ATP) synthesis.* I: complex I (NADH dehydrogenase); II: complex II (succinate dehydrogenase); III: complex III; IV: complex IV (cytochrome *c* oxidase); *b*: cytochrome *b*; c_1: cytochrome c_1; *c*: cytochrome *c*; *a*: cytochrome *a*; a_3: cytochrome a_3; FeS: iron sulfur center; Cu: cuprous protein; Q: ubiquinone (coenzyme Q); F0: ATP synthase (F_0 subunit); F1: F_1 subunit; PT: phosphate ion transporter. *Source: Modified from Fig. II-90 in Kagawa and Nozawa, Illustrated Medical Chemistry (2001, p. 212).*

and promotes aging. Reduced complex I in the electron transfer system induces Alzheimer disease. Deficiency of complexes I and III may induce acidosis.

Glycogen Metabolism

Glycogen Synthesis

Glycogen is the reserve polysaccharide in the body and is mainly comprised of hepatic glycogen. Glycogen is synthesized in the liver and muscles. α-D-Glucose combines to form glycogen continuously. There is a reduced sugar that indicates reduction characteristics, and many non-reducing residues that do not indicate reduction in the glycogen molecule. When glucose binds to glycogen, or when it is cleaved, a non-reducing residue is involved. The production of glycogen from glucose is shown in Figure 4.4. An important enzyme in glycogen synthesis is glycogen synthase. By the action of this enzyme, one glucose can combine with glycogen. However, when glucose binds to glycogen, it links with a non-reducing residue.

Glycogen Degradation

Glycogen phosphorylase acts on the reaction at the initiation of glycogen degradation (Figure 4.5). Thereby, glucose can be obtained from glycogen. Glycogen phosphorylase causes phosphoroclastic cleavage into glycogen, and produces glycogen-1-phosphate. However, a non-reducing terminal is removed when cleaving glucose from glycogen. Moreover, if glucose-1-phosphate produced from glycogen is changed into G6P by phosphoglucomutase, it can proceed directly to glycolysis. It should be mentioned that the direct progress to glycolysis leads to the exclusion of ATP consumption required for converting glucose to G6P. Therefore, three ATPs are made if the glucose generated from glycogen is used by glycolysis. The glycogen metabolism is controlled by the activity of glycogen synthase and glycogen phosphorylase. The major regulatory feature involved in the metabolism is phosphorylation, which inactivates glycogen synthase and activates glycogen phosphorylase.

Use of Glucose-6-Phosphoric Acid

Glycogen exists mainly in the liver and muscles. When the glycogen in the liver is broken down, glucose is produced and impacts maintenance of the blood sugar level, typically at 80–100 mg/dl. If muscle glycogen is broken down, it proceeds to glycolysis and becomes an energy source in the muscle cells. The difference is based on the enzymatic activity of glucose-6-phosphatase.

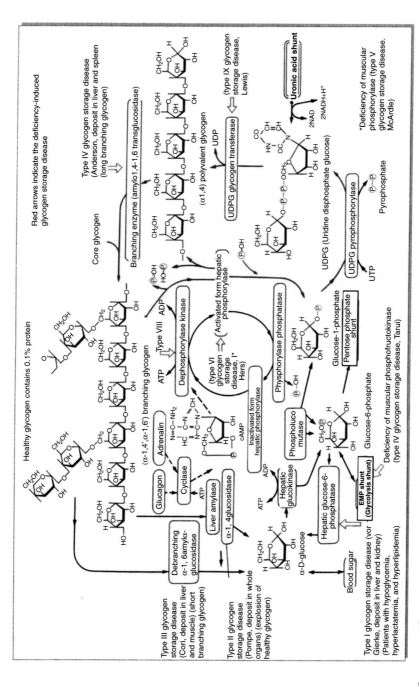

Figure 4.4 Glycogen shunt and activation of phosphorylase by cyclic adenosine monophosphate (cAMP). Source: Modified from Fig. II-15A in Kagawa and Nozawa, Illustrated Medical Chemistry (2001, p. 132).

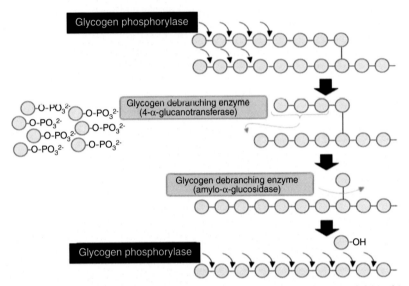

Figure 4.5 *Glycogen hydrolysis by glycogen phosphorylase and glycogen debranching enzyme.*

Although glucose-6-phosphatase participates in the conversion of glucose-6-phosphoric acid to glucose, glucose-6-phosphatase activity is strong in the liver and kidney because of its high production. By contrast, muscles have low glucose-6-phosphatase activity. The muscular G6P produced is therefore supplied only to energy production, whereas the G6P in the liver goes into both glycolysis and production of glucose, which is used for maintaining the blood sugar level.

Glyconeogenesis

Glyconeogenesis is a shunt for the synthesis of sugars such as glucose and glycogen from substances other than sugars. An example is the conversion from lactic acid to glucose. It passes through the following process, from the lactic acid elevated by a glycolytic shunt to make glucose by glyconeogenesis: lactic acid is converted into alanine and then carried into the liver. Alanine in the liver is changed back to lactic acid and synthesized to glucose by glyconeogenesis. However, glyconeogenesis occurs partly in the kidney. If enzymes are involved in the reaction of the glycolytic shunt, it is possible to make glucose by glyconeogenesis except for three one-way (irreversible) reactions in the glycolytic shunt. The three reactions are: (1) conversion of glucose into G6P by hexokinase; (2) conversion of F6P into F1,6BP by

PFK-1; and (3) conversion of PEP into pyruvate by pyruvate kinase. The reverse reactions of (1) and (2) are solved by the utilization of other enzymes, glucose-6-phosphatase and fructose-1,6-bisphosphatase, respectively. However, the last reaction (3) is more complex. First, pyruvate is carried into mitochondria and then converted into oxaloacetate by the catalytic action of pyruvate carboxylase and using CO_2 and ATP. Next, the generated oxaloacetic acid is changed to malate, which is carried out of mitochondria through the malate shuttle and then regenerated to oxaloacetate. Finally, PEP is formed from oxaloacetate by the action of phosphoenolpyruvate carboxylase. Glucose is produced from two molecules of lactic acid in this reaction. In glyconeogenesis, six ATP molecules are needed to generate one molecule of glucose.

Pentose Phosphate Pathway

The pentose phosphate shunt generates the reduced form of nicotinamide adenine dinucleotide phosphate (NADPH) and D-ribose-5-phosphoric acid (Figure 4.6). NADPH is a coenzyme which acts in the cytoplasm and is linked with fatty acid synthesis. D-Ribose-5-phosphate is associated with the synthesis of nucleic acids (DNA and RNA).

Blood Groups

The A blood group has a gene (A-transferase) which expresses antigen A; a gene in the AB blood group expresses both the A and B antigens for the genes (A-transferase and B-transferase) (Figure 4.7); and in the B blood group, the B-transferase gene expresses antigen B. Antigens A and B are chemically arranged from antigen H by H-transferase. The gene that determines an ABO blood group is located on chromosome 9. The gene coding for antigen H is located on chromosome 19, and transforms the H-transferase precursor to antigen H. This gene is not expressed in people with the Bombay blood type. Besides classification of the ABO blood group system, there are other blood group systems based on Rhesus factor, Lewis and MN blood groups. The distribution of ABO blood groups differs greatly by area and ethnicity; for example, about 40% of Japanese people have the A blood group, 30% O, 20% B and 10% AB, whereas 90% or more of native South Americans are blood group O, and this figure may exceed 99% depending on the area. A relationship between the ABO blood group and illness was noticed in 1980. An outline was published in the scientific journal *Nature* in 2000, and completed by the Human Genome Project (International Human Genome Sequencing Consortium, 2000). A weak

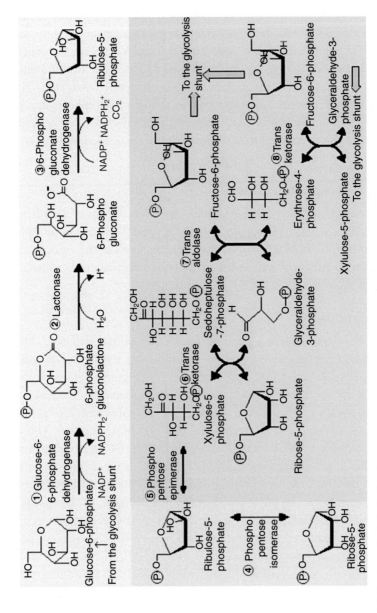

Figure 4.6 Pentose phosphate shunt.

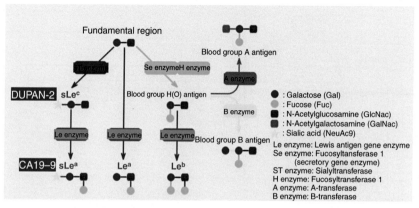

Figure 4.7 ABH and Lewis blood groups.

correlation was found between ABO blood group and *Helicobacter pylori* infection. In contrast, there is a close relation between the Lewis secretory form and *H. pylori* infection. Bone marrow transplantation can change a recipient's ABO blood group. For example, the ABO blood group of the late Ichikawa Danjuro XII, who had myelocytic leukemia, changed following bone marrow transplantation.

Glycoprotein Synthesis

The first type of glycoprotein is arranged post-translationally to an asparagine-linked mannose-rich sugar chain on the endoplasmic reticulum, and then the glycoprotein can be arranged into a complex sugar or an O-linked sugar chain in the Golgi area. The second type of glycoprotein has a kind of a glycolipid called glycosylphosphatidylinositol (GPI), which is also arranged post-translationally on the endoplasmic reticulum (Figure 4.8). The latter GPI glycoproteins exist in a cell membrane that is linked by the GPI anchor. There are at least 150 kinds of GPI protein, including alkaline phosphatase, prion glycoprotein, cholinesterase and Tamm–Horsfall protein localized in the renal distal tubules. The synthesis of GPI and details of the mechanisms of GPI and protein linkage are still obscure, and the physiological role of the GPI anchor is also still unknown.

The post-translational modification of a GPI anchor glycoprotein can be synthesized by GPI-dependent amidotransferase. The precursor of GPI glycoprotein has a certain hydrophobic amino acid sequence that

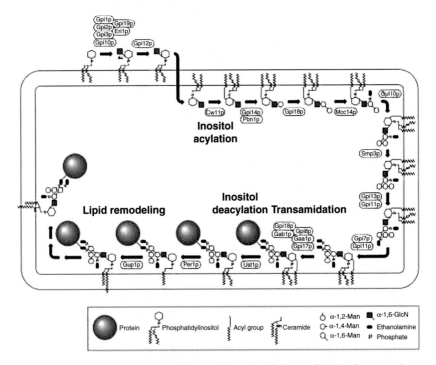

Figure 4.8 *Synthesis mechanism of glycosylphosphatidylinositol (GPI)-glycoprotein.*

is cleaved off and replaced by the GPI anchor, possibly by an amido-transferase, although this enzyme has not yet been discovered. The abnormal degradation steps in the early stages of the GPI constructional system in paroxysmal hemoglobinuria (PNH or red diaper syndrome) have a critical effect on the complement fixation reaction at night. PNH is an acquired disease in which the membrane of red cells is destroyed by complement-mediated hemolysis. Destruction of red cells takes place within the blood vessel, and this is called intravascular hemolysis. The complement response in connection to this disease occurs via a protein called complement C3. The disease controls the quality of the membrane protein, such as CD34, CD55 and CD59. The protein-linked GPI of the cell membrane in this disease is not synthesized or there is a lack of GPI anchors. For this reason, control of C3 activation is lost and the red cell membrane is destroyed. Abnormalities of the phosphatidylinositol glycan class A gene (*PIG-A*) in the synthesis of GPI are also involved in PNH.

LIPID METABOLISM

β-Oxidation is a metabolic pathway that oxidizes fatty acids, and generates fatty acyl-CoA (a thioester of fatty acid and CoA) and acetyl-CoA. β-Oxidation consists of a series of four repeated reactions, in which a molecule of acetyl-CoA is generated, and an end-product of the fatty acid by β-oxidation is also acetyl-CoA. Since oxidation at the β position of the fatty acyl-CoA was performed stepwise, it was named β-oxidation. β-Oxidation is the first of three stages (β-oxidation, TCA cycle and electron transfer system) in the metabolism of fatty acids. The generated acetyl-CoA is linked to the TCA cycle and acetyl-CoA oxidizes to CO_2. β-Oxidation takes place not only in mitochondria but also in peroxisomes. Mitochondria metabolize short-chain fatty acids and peroxisomes long-chain fatty acids. The glycerol or fatty acid in the body may remove energy, and glycerol is stored in adipose cells as an ester. Since triglycerides are accumulated as lipid droplets in the cells, the storage lipids do not increase the osmotic pressure of cytoplasm and do not hydrate. To use triglycerides for energy production, they are carried as lipid droplets to the organs (skeletal muscle, heart, renal cortex, liver, small intestine, etc.) as follows. First, hormone-sensitive lipase moves on to the surface of a lipid droplet. Lipase can hydrolyze triglycerides. The hydrolyzed fatty acid flows in the bloodstream and is carried by the muscle tissue. The fatty acid isolated from serum albumin is taken into cells through the fatty acid transporter. Thus, after being taken into the cells, acetyl-CoA is generated via activation of fatty acid and β-oxidation. So that the fatty acid taken into the cell can achieve stable C–C binding, it is activated by the enzyme acyl-CoA synthase (also called fatty acid thiokinase), which exists in the cytoplasmic side of the mitochondrial outer membrane, in the following reaction:

$$\text{Fatty acid} + \text{CoA} + \text{ATP} \rightarrow \text{Fatty acyl-CoA} + \text{AMP} + \text{PP}_i$$

This reaction consists of two steps. Initially, a reaction of the carboxylate ion of fatty acid with the phosphate (β, γ-phosphate) of ATP forms fatty acyl adenylate and pyrophosphate (PP_i). Next, the thiol residue of CoA moves the nucleophilic attack of the carbon of the acyl residue, and generates fatty acyl-CoA and adenosine monophosphate (AMP).

Fatty acyl-CoA is a high-energy compound. Because acyl-CoA is capable of penetrating the mitochondrial inner membrane directly, carnitine plays an important role in the fatty acyl shuttle. Fatty acyl-CoA binds temporarily with carnitine and then generates fatty acid acyl carnitine. This

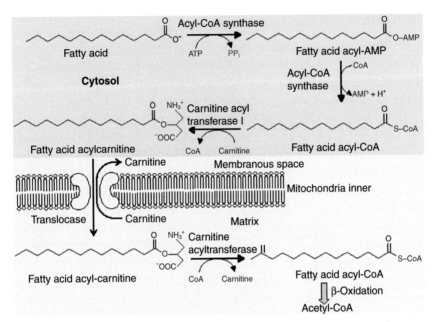

Figure 4.9 *Uptake of fatty acid in mitochondria.* CoA: coenzyme A; AMP: adenosine monophosphate.

reaction is catalyzed by carnitine acyltransferase I, embedded in the mitochondrial outer membrane (Figure 4.9). Although the fatty acid acyl carnitine is thought to be generated in the intermembrane space and cytosol side of the outer membrane in mitochondria, it is still unclear where the carnitine adducts are generated. Fatty acid acyl carnitine crosses the membrane through facilitated diffusion by the acyl carnitine/carnitine transporter, and shifts in the matrix. Then, the fatty acyl group is transferred from carnitine to CoA by an enzyme located in the inner mitochondrial membrane, carnitine acyltransferase II, to yield acyl-CoA. Free carnitine moves to the intramembrane space again via the acyl carnitine/carnitine transporter. This fatty acyl-CoA transportation system is called the carnitine shuttle. The fatty acyl-CoA that is carried into the mitochondria undergoes oxidation with the enzymes in the matrix.

As mentioned above, the β-oxidation reaction consists of four repeated reaction steps, which remove two carbons from the carboxy-terminus of the fatty acyl chain as an acetyl-CoA. For example, palmitic acid (C16), which has an even number of carbon atoms, is converted into palmitoyl-CoA, which is oxidized seven times within the mitochondria and produces seven molecules of acetyl-CoA (CH_3CO-S-CoA). Furthermore, since two carbons are left behind at the unreduced end as acetyl–CoA, eight molecules of acetyl–CoA are produced in the whole reaction. Odd-carbon fatty acids undergo a similar reaction and produce propionyl–CoA (CH_3CH_2CO-S-CoA), consisting of three carbons, as well as the acetyl–CoA.

The β-oxidation enzyme system requires four enzymes and the above-mentioned acyl–CoA synthase of β-oxidation. These are acyl–CoA synthase (inside chain EC 6.2.1.2, long chain EC 6.2.1.3), acyl–CoA dehydrogenase (EC 1.3.99.3), enoyl–CoA hydratase (EC 4.2.1.17) and β-hydroxyacyl–CoA dehydrogenase (EC 1.1.1.35). The stages of β-oxidation catalyzed by these enzymes are described in detail. Although the following reaction occurs only in the oxidation direction *in vivo*, it is reversible *in vitro*.

The first phase is the oxidation reaction catalyzed by acyl–CoA dehydrogenase. In this step, a double bond is formed between α-carbon and β-carbon, and *trans*-Δ^2-enoyl–CoA is produced. The Δ symbol indicates the position of a double bond. Four isozymes of acyl–CoA dehydrogenase exist in mammals, and act specifically on the acyl chain of the short chain, the inside chain, the long chain and the super-long chain. An electron moves to flavine adenine dinucleotide (FAD), which is a prosthetic residue of these enzymes, by formation of a double bond, and it is immediately sent to the mitochondrial electron transport chain. Another FAD bound to the hydrophilic electron transfer flavoprotein (ETF) catches this electron.

In the second step, a reaction catalyzed by enoyl–CoA hydratase adds H_2O to the double bond formed in the preceding step, forming β-hydroxyacyl–CoA (β-hydroxyacyl–CoA and 3-hydroxyacyl–CoA). This reaction progresses in a stereoselective manner, forming only the L-type.

The third step is catalyzed by β-hydroxyacyl–CoA dehydrogenase and produces β-ketoacyl–CoA. This enzyme acts only on L-type β-hydroxyacyl–CoA. This reaction is NAD$^+$ dependent. NAD$^+$ accepts an electron to produce NADH and the electron is further passed to complex I of the electron transfer system. The first three steps are the reaction for destabilizing the stable C–C bond.

The fourth step is the reaction due to β-ketoacyl–CoA thiolase, which also called thiorase, β-ketothiolase and acetyl–CoA acetyltransferase. The enzyme catalyzes a thiolytic cleavage (thiolysis) of β-ketoacyl–CoA and CoA, resulting in the two carbon-shortened fatty acyl–CoA and acetyl–CoA. The α-β carbons are cleaved because the thiol residue (–SH) of CoA attacks the carbonyl carbon of β-ketoacyl–CoA nucleophilically.

The fatty acyl–CoA with the shortened carbon chain serves as a substrate for the first step in the following β-oxidation, and the reaction is repeated until all parts of the fatty acyl chain have oxidized to acetyl–CoA (or propinyl–CoA). The equation can be described using palmitate as an example. β–Oxidation produces a large amount of acetyl–CoA as the end-product, and acetyl–CoA is a flexible substance used in various metabolic systems. Acetyl–CoA is utilized for the following reactions.

Eicosanoids, such as prostaglandins (PGE_2, $PGF_{2\alpha}$ and PGD_2) and leukotrienes (LTB_4), are produced from an unsaturated fatty acid, arachidonic acid. They are involved in birth, uterine contractions, sleep and production of reactive oxygen species (ROS). Edaravone is a scavenger of ROS and induces sleep via the circadian rhythm; this is also adjusted by clock genes such as *Bmal1*. Melatonin may be taken to prevent jet lag on an overseas trip.

Steroid Hormones

Cytochrome cholesterol side-chain cleavage enzyme (P450scc) catalyzes the synthesis from cholesterol of sex hormones and other steroid hormones, e.g. estrogen, testosterone, aldosterone and cortisone (Table 4.1). Loss of P450scc activity may result in hermaphroditism, decreased estradiol may lead to osteoporosis, and reduced testosterone may induce cryptorchidism and sterility in males. Cortisone acts on skeletal metabolism and glyconeogenesis, and has an anti-inflammatory effect. The molecule angiotensin II is formed from angiotensin I by angiotensin-converting enzyme (ACE) produced in the kidney, and promotes secretion of a mineralocorticoid aldosterone, which has a blood pressure elevating effect. The elevation of aldosterone can induce diabetes and acidosis. Adrenal cortex hormone corticosteroid is involved in a wide range of physiological systems, such

Table 4.1 Synthesis of sex hormones

Cholesterol		
↓ CYP11A1	3β–HSD	
Pregnenolone	⟶	Progesterone
↓ CYP17	3β–HSD	↓ CYP17
17α-Hydroxypregnenolone	⟶	17α-Hydroxyprogesterone
↓ CYP19	3β–HSD	↓ CYP19
Dehydroepiandrosterone (DHEA)	⟶	Androstenedione
↓ 3β–HSD	17β–HSD	↓ 17β–HSD
Androstenedione	⟶	Testosterone
↓ CYP19 (aromatase)	17β–HSD	↓ CYP19 (aromatase)
Estrone	⟶	17β-Estradiol

CYP: 450 families (CYP2H2 of alcohol lysis enzyme, etc.); HSD: hydroxysteroid dehydrogenase isomerase.

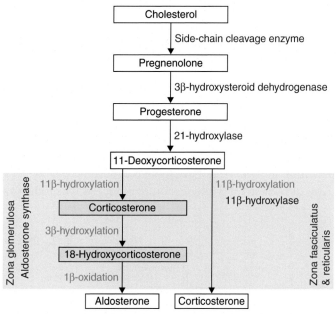

Figure 4.10 *Synthesis of adrenal cortex hormones.* Blue: enzyme; green: action performed; red: location. *Source: Modified from Wikipedia.*

as control of inflammation, carbohydrate metabolism, protein catabolism, blood electrolytes and immunoreactivity. The hormones are secreted in response to various factors such as stress and invasion, and are also used in medicines. Adrenal cortex hormones consist of glucocorticoids and mineralocorticoids. Glucocorticoids function as anti–inflammatory medicines and decrease acidophilic material by controlling carbohydrate, fat and protein metabolism and preventing the generation of phosphatide. Mineralocorticoids are mainly produced in the adrenal gland, and act to control the balance of electrolytes and water. The typical hormones are cortisol and aldosterone (Figure 4.10). There are several receptors (R) for steroid hormones (S) on the biomembrane, and the SR complex or the resulting product binds to a specific nucleic acid sequence in the nucleus. Then, a particular protein can be produced automatically.

The medicines ezetimibe and mevalotin, as statin derivatives, specifically suppress the cholesterol transporter and 3-hydroxy-3-methyl-glutaryl-CoA (HMG-CoA) reductase. Cholesterol is required for normal cell function in mammalian cell membranes. In synthesis of the lipid, acetyl-CoA is used as the starting material in the intracellular endoplastic reticulum of cells. Cholesterol is carried into low–density lipoprotein (LDL) as the acyl ester of cholesterol in the blood and the lipoprotein cholesterol is incorporated into cells through receptor-mediated endocytosis, by inclusion into clathrin-coated pits and subsequent degradation by lysosomes. Cholesterol may be divided into exogenous cholesterol, which is formed from bile acid and dietary sources and absorbed from the intestinal duct, and endogenous cholesterol, which is mainly synthesized via mevalonic acid and squalene from acetyl-CoA in the liver. The amount of biosynthesis is adjusted to allow for changes in the amount of exogenous cholesterol. The total amount of cholesterol in the body is about 35 g for a person with a body weight of 68 kg. Although most of the cholesterol is present in the cell membrane, some is circulating (Figure 4.11). About 800 mg/day of cholesterol is endogenously produced when low–cholesterol

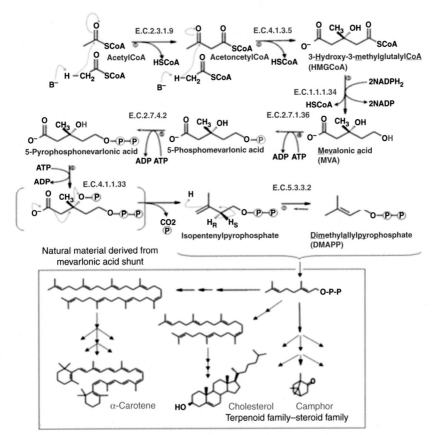

Figure 4.11 *Cholesterol metabolism.* EC 2.3.1.9: acetyl-CoA-acetyltransferase (thiolase); EC 4.1.3.6: HMG-CoA synthase; EC 1.1.1.34: HMG-CoA reductase; EC 2.7.1.36: mevalonate kinase; EC 2.7.4.2: 5-phosphomevalonate kinase; EC 4.1.1.33: diphosphomevalonate decarboxylase; EC 5.3.3.2: isopenthenyldiphosphate δ-isomerase.

food is ingested, and 20–25% of the circulating cholesterol is made in the liver. This implies that 50% of cholesterol in the body steroids exists in the bloodstream. The level of cholesterol is the highest in the liver, followed by the skin; it has not been reported in other organs except for the hippocampal region of the brain. In mammals including humans, steroid hormones needed in organs other than the skin and liver are not biosynthesized.

Reduced production of testosterone has been shown to lead to schizophrenia via reduced NADH metabolism (Liu et al., 2011).

Cholesterol is produced by *de novo* synthesis and transported using the phospholipid complex containing its esters from the liver to the plasma. When cholesterol is stored in the body, it does not have a special form. For example, the lipoprotein (LDL-cholesterol) in the middle of transportation also plays a role as a reservoir of cholesterol in the body. The cholesterol ester used in peripheral organs with phospholipids returns cholesterol in response to hydrolysis by lysosomes. Since cholesterol has such a dynamic state, and while biosynthesis of cholesterol by absorption from a meal and in the liver is indispensable, the cholesterolemia caused by ingesting excess cholesterol can create problems. Hypercholesterolemia is caused not only by increased exogenous cholesterol from a meal but also by a reduction in LDL-cholesterol receptors in peripheral organs. In familial hypercholesterolemia (FH), cholesterol uptake is decreased as a result. It is also thought that the lipid metabolism is abnormal. In addition, lifestyle diseases such as metabolic syndrome, which is characterized by increased waist circumference, blood pressure, glycosylated hemoglobin (HbA_{1c}) and high-density lipoprotein (HDL)-cholesterol, are attracting attention as complex metabolic disorders of sugars and lipids. MIYAIRI 588 (Miyarisan Pharmaceutical Co.) has been used to treat high fatty acid levels in mice with non–alcoholic fatty liver disease (Seo et al., 2013).

Lipoprotein Metabolism

Lipoproteins exist in the blood plasma. To exist stably in the plasma, lipids, except for polar molecules such as free fatty acids, must bind with protein as apoprotein. Lipoproteins are spherical particles containing the many triglycerides and cholesterol necessary to support cells.

Lipoprotein Species

Lipoprotein species and their major apolipoproteins are shown in Table 4.2. Lipoproteins can be classified according to electrophoretic or supercentrifugal methods. Lipoproteins can be classified as chylomicrons, very low-density

Table 4.2 Lipoprotein species and their major apolipoproteins

Apolipoprotein	Amino acid residue	Biosynthetic organ(s)	Lipoprotein species	Functions
A-I	243	Liver, intestine	HDL3, HDL2 CM	Activation of LCAT HDL synthesis
A-II	77x2	Liver, intestine	HDL3, HDL2	HDL metabolism; homodimer
A-IV	376	Liver, intestine	CM	Activation of LCAT
B-100	4536	Liver	VLDL, LDL	Bind to LDL receptor; structural protein for VLDL or LDL
B-48	2152	Intestine	CM	Structural protein for CM
C-I	57	Liver	CM, VLDL	Induction of atherosclerosis
C-II	79	Liver	CM, VLDL	Activation of LPL
C-II	299	Liver	CM, VLDL	Induction of atherosclerosis
E	299	Liver	CM, VLDL	Bind to apoE receptor; bind to LDL receptor

HDL: high-density lipoprotein; CM: chylomicron; VLDL: very low-density lipoprotein; LDL: low-density lipoprotein; LCAT: lecithin cholesterol acyltransferase; LPL: lipoprotein lipase; apoE: apolipoprotein E.

lipoprotein (VLDL), intermediate-density lipoprotein (IDL), low-density lipoprotein (LDL), high-density lipoprotein (HDL) and very high-density lipoprotein (VHDL), and the standard profile is established by the World Health Organization (WHO) classification of hyperlipidemia.

IDLs can be detected in lipoprotein disorders. When triglyceride in a particle of VLDL is degraded and lost, the ratio of cholesterol is increased, and for LDL-cholesterol is called remnant lipoprotein, in accordance with that in which chylomicron loses triglyceride; these are equivalent to the IDL fraction. Small dense low-density lipoprotein (sd-LDL) is a triglyceride-rich LDL molecule. IDLs in patients with advanced arteriosclerosis can be detected by supercentrifugal or disc electrophoresis methods. IDL appears as a band between VLDL and LDL (β and pre-β). Chylomicrons are lipoproteins with a density less than 0.96 g/ml, and cholesterol and triglycerides are contained at a rate of about 1:10 in chylomicrons. The lipid absorbed from the intestinal duct is reconstructed by lipoprotein on the proximal intestinal membrane, and carried into the liver through the lymph tube. The lipoprotein that acts as the lipid carrier is chylomicron, whose major apoprotein is ApoB-48. LPL deficiency leads to remarkably high chylomicronemia. Abetalipoprotein-emia is elicited by the loss of microsome triglyceride transfer protein (MTP), which is needed for lipoprotein synthesis, and its major symptoms are night blindness and peripheral nerve paralysis owing to a reduction in carrier particles of lipophilic vitamins A and E. IDL, a lipoprotein of density 1.006– 1.019 g/ml, is generated by hydrolyzing VLDL and chylomicron by LPL and is also called a remnant. In general, patients with metabolic syndrome have a background of insulin resistance. LPL activity is reduced in patients with type III hyperlipemia and the ApoE variant; the circulating IDL is not capable of binding to receptors for LDL, VLDL and a lipoprotein receptor-related protein (LRP), resulting in accumulation in the bloodstream.

Using the polyacrylamide gel electrophoresis method, IDL can be identified qualitatively as a midband (Figure 4.12). Moreover, the amount

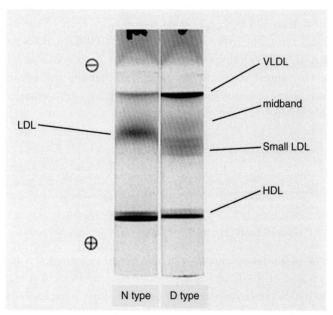

Figure 4.12 *Species of lipoproteins.* LDL: low-density lipoprotein; VLDL: very low-density lipoprotein; HDL: high-density lipoprotein; N type: healthy subject; D type: patient with diabetes mellitus; Midband: intermediate-density lipoprotein; Small LDL: small dense low-density lipoprotein.

can be determined quantitatively by the remnant-like particle cholesterol (RLP-C) measurement kit using an anti-ApoA-I antibody and anti-ApoB-100 antibody. Evaluation based on quantification of ApoB-48 has been proposed. LDL has a density of 1.006–1.063 g/ml and a particle size of 26–27 nm. LDL is sometimes called "bad cholesterol." Through the LDL receptor that recognizes ApoB and ApoE, the lipoprotein is mainly taken into the liver and metabolized. LDL receptor deficiency is called familial hypercholesterolemia (FH); the total cholesterol concentration in FH homozygotes is at least 600 mg/dl, even in adolescence, causing severe arteriosclerosis such as ischemic heart disease. When LDL is modified by oxidation, denaturing and glycation, it loses its affinity to the LDL receptor. In that case, oxidized LDL is taken into the macrophage via the scavenger receptor BI (LOX-1), resulting in the formation of fat-laden foam cells and thick atheromatous plaque.

Oxidative modification changes LDL and HDL into smaller particles, and therefore the blood vessel wall plays a role in ischemic heart disease (Matsunaga et al., 2013). Patients with hyperlipidemia have increased small dense LDL (sd-LDL) and sd-HDL in the bloodstream, but these triglyceride-rich particles can be reduced by fibrate-based medicine. In healthy HDL, the lipoprotein density is 1.063–1.21 g/ml and the diameter of the HDL particle is 10–12.5 nm. The major apoproteins in the HDL particles are ApoA-I and ApoA-II. HDL-cholesterol, also known as "good cholesterol," cleans cholesterol accumulated on the vascular endothelium and is thought to protect from arteriosclerosis through the reverse cholesterol transfer system. Reverse cholesterol transfer protein (RCTP) facilitates the exchange of neutral fat between VLDL and HDL, and cholesteryl ester transfer protein (CETP) exchanges cholesterol between LDL and VLDL. All the cholesterol can return to the liver by RCTP (Figure 4.13). However, oxidized HDL seems to lose its beneficial characteristics.

AMINO ACID METABOLISM

The transfer of amino residues is catalyzed by transaminases. Amino acids and aspartic acid of the TCA cycle produce oxaloacetic acid (Figure 4.14). After being hydrolyzed by asparaginase and changing asparagine into aspartic acid, it is similarly decomposed into oxaloacetic acid. Arginine, glutamine, histidine and proline are decomposed into glutamic acid, and glutamic acid is oxidized to 2-oxoglutaric acid by glutamic acid dehydrogenases (Figure 4.15). The reaction is hydrolyzed by glutaminase and glutamine is

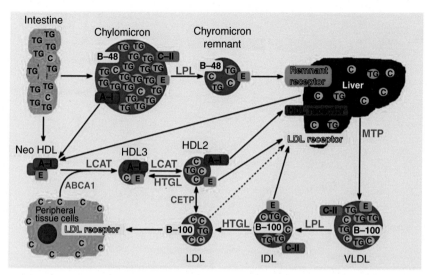

Figure 4.13 *Metabolism of lipoproteins.* TG: triglyceride (neutral lipid); C: cholesterol; LPL: lipoprotein lipase; E: apoprotein E; HDL: high-density lipoprotein; LDL: low-density lipoprotein; IDL: intermediate-density lipoprotein; VLDL: very low-density lipoprotein; MTP: microsome triglyceride transfer protein; LCAT: lecithin cholesterol acyltransferase; HTGL: hepatic triglyceride lipase; ABCA1: ATP binding cassette A1; CETP: cholesteryl ester transfer protein.

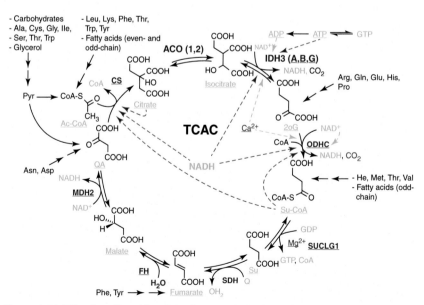

Figure 4.14 *Tricarboxylic acid cycle and amino acids. Source: Modified from Wikipedia.*

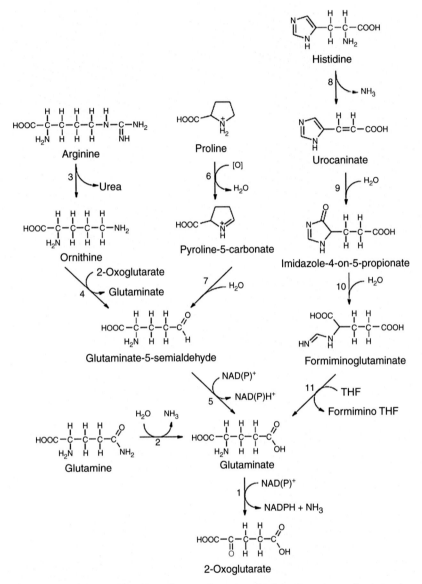

Figure 4.15 *Metabolism of histidine.* 1: glutamic acid dehydrogenases; 2: glutaminase; 3: arginase; 4: ornithine-5-transaminase; 5: glutamate semialdehyde dehydrogenase; 6: proline oxidase; 7: non-enzymatic reaction; 8: histidine ammonia lyase; 9: urocanate hydratase; 10: imidazolone propionase; 11: glutamic acid formiminotransferase.

changed into glutamic acid. Non-oxidative deamination of histidine opens the imidazole ring, producing N-formiminoglutamic acid, and transfer of the formimino residue to tetrahydrofolic acid (THF) produces glutamic acid and 5-hormuimino tetrahydrofolic acid (5-hormimino THF). Glutamic acid produces arginine and proline via glutamic acid 5-semialdehyde. Methionine reacts with ATP and forms S-adenosylmethionine (SAM or AdoMet) (Figure 4.16). SAM is a strong methyl residue donor, which converts phosphatidylethanolamine and epinephrine (adrenaline) into phosphatidylcholine and norepinephrine (noradrenaline), respectively. If a substrate is supplied with a methyl residue, SAM will serve as S-adenosylhomocysteine. S-Adenosylhomocysteine is hydrolyzed into adenosine and homocysteine, which is further modified into methionine (by reproduction of methyl residue) or cystathionine (by binding to serine).

High homocysteinemia is a disorder induced by an imbalance between the generation and decomposition of methylhomocysteine. The onset of high homocysteinemia in the neonate causes cardiovascular disease, neural tube defects and anencephaly. Genetic homocysteinemia in infants can be prevented by the ingestion of vitamin B6, B_{12} and folic acid during pregnancy.

Branched-Chain Amino Acid Metabolism

The branched-chain amino acids are leucine, isoleucine and valine.

Leucine

A lack of branched-chain α-keto acid dehydrogenase (BCKDH) induces maple syrup urine disease (MSUD). Excess amounts of the branched α-keto acid, due to deficiency of the enzyme, are excreted in the urine, producing a smell similar to maple syrup. MSUD is an autosomal recessive disease, and if it is not treated immediately after diagnosis the patient will soon die.

Isoleucine and Valine

The first four steps in the decomposition of isoleucine are catalyzed by four enzymes (branched amino acid transaminase, branched α-keto acid dehydrogenase, acyl-CoA dehydrogenase and enoyl-CoA hydratase), which also metabolize valine in the same manner (Figure 4.17). In the degradation pathway of 3-hydroxy-2-methylbutyryl-CoA derived from isoleucine, acetyl-CoA is generated besides propionyl-CoA (Figure 4.18).

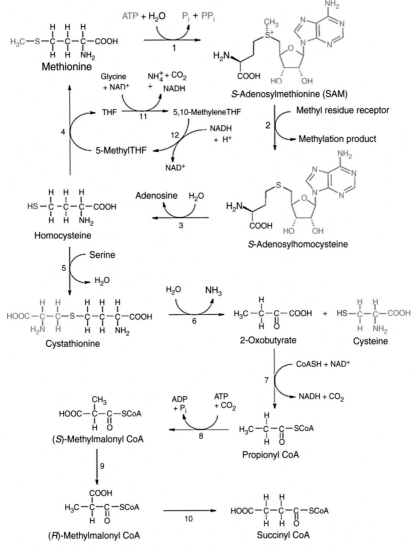

Figure 4.16 *Metabolism of methionine.* 1: methionine adenosyltransferase; 2: methyl transferase; 3: adenovirusyl homocysteinase; 4: 5-methyl THF-homocysteine methyl transferase (methionine synthase); 5: cystathionine β-synthase (PLP enzyme); 6: cystathionine γ-lyase (PLP enzyme); 7: α-keto acid dehydrogenase; 8: propionyl-CoA carboxylase; 9: methylmalonyl-CoA epimerase; 10: methylmalonyl-CoA mutase; 11: glycine cleavage system; 12: 5,10-methylene tetrahydrofolate reductase; CoA: coenzyme A.

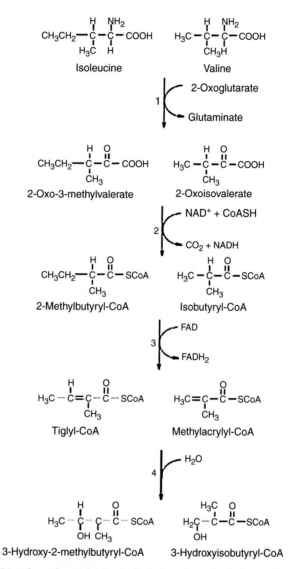

Figure 4.17 *Disorders of methionine hydrolysis.* 1: branched-chain amino acid transaminase; 2: branched-chain α-keto acid dehydrogenase (BCKDH); 3: acyl-CoA dehydrogenase; 4: enoyl-CoA hydratase; CoA: coenzyme A.

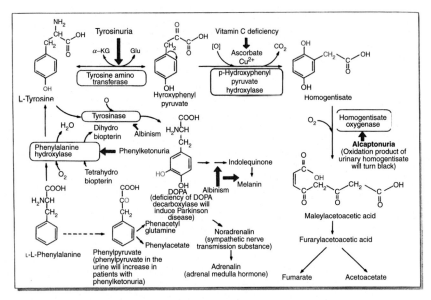

Figure 4.18 *Metabolism of phenylalanine and tyrosine.* Source: Modified from Kagawa and Nozawa, *Illustrated Medical Chemistry (2001, p. 168).*

Phenylalanine and Tyrosine

Phenylalanine is metabolized into acetoacetic acid and fumaric acid via tyrosine. A tyrosine metabolite, DOPA, is converted into the neurotransmitters epinephrine and norepinephrine. A lack of these neurotransmitters is a causal factor in Parkinson disease and schizophrenia. Administration of the β-blocker chlorpromazine is effective in their treatment, but can cause osteoporosis as a side-effect. When patients with schizophrenia are treated with chlorpromazine, the physician should advise taking active vitamin D.

Patients with alcaptonuria have black urine due to the oxidation of homogentisic acid. Phenylketonuria is induced by an error in phenylalanine metabolism, which is due to the lack or dysfunction of phenylalanine hydroxylase, involved in the conversion into tyrosine. By this metabolism, the concentration of phenylalanine in the blood is elevated. A patient with phenylalaninemia (indicative of high levels of phenylalanine) produces higher concentrations of phenylpyruvic acid, phenyllactic acid and phenylacetic acid, resulting in phenylketonuria and schizophrenia.

Tyrosinase deficiency can cause albinism with a lack of the neurotransmitter dopamine, resulting in symptoms of schizophrenia.

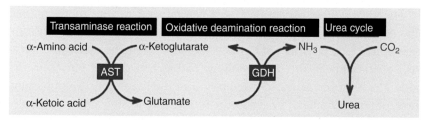

Figure 4.19 *Urea cycle.* AST: aspartate aminotransferase; GDH: glutamate dehydrogenase.

UREA CYCLE

In the urea cycle (also called the ornithine cycle), urea is generated from ammonia (Figure 4.19). This metabolism was discovered by Hans Krebs and Kurt Henseleit in 1932. This cycle occurs in almost all land vertebrates, and the reactions in the cycle take place in the mitochondria of hepatic cells and cytosol.

NUCLEIC ACID METABOLISM

Purine Metabolism

The purine bases are synthesized from glutamine, glycine, etc., based on the novel synthesis shunt *de novo* cycle from 5-phosphoribosyl-1-pyrophosphate (PRPP). This leads to inosine monophosphate (IMP), which serves as a precursor of adenine or guanine. An amino residue is added from glutamic acid by PRPP amidotransferase.

Nucleotide Decomposition

As a result of deamination by adenosine deaminase, adenosine is changed into inosine, which is further converted into hypoxanthine through dephosphorylation and subsequent deribosization. Oxidation of hypoxanthine yields uric acid. In general, the solubility of uric acid in body fluid is 7.0 mg/dl. Gout is caused when the blood concentration exceeds this value.

Gout

Gout is a metabolic syndrome. In this illness, uric acid accumulates in the organs owing to a defect in purine metabolism. Uric acid may be stored in the joints. People should not consume a lot of beef, pork and chicken. Draft beer contains ATP, and the resulting AMP and IMP lead to increased concentrations of uric acid.

Hereditary gout is classified as hypersynthetic or due to the partial absence of hypoxanthine–guanine phosphoribosyltransferase (HGPRT), known as Lesch–Nyhan syndrome or Kelley–Seegmiller syndrome. HG-PRT participates in the purine salvage pathway to recover purine in the body. The X-linked recessive hereditary gout is related to increased biosynthesis of purine bodies. The syndrome is seen only in boys and is characterized by mental problems and self-injurious behavior. Although the urate synthesis inhibitory medicine urinome has been used to treat gout, urinome may downregulate uric acid reabsorption in the renal tubules via peroxisome proliferator-activated receptor-α (PPARα) (Kunishima et al., 2007).

Adenosine phosphoribosyltransferase (APRT) changes adenine into AMP. In APRT deficiency, a rare autosomal recessive inheritance disease, purine salvage shunt urolithiasis and renal insufficiency by DHA are seen.

In acute gouty arthritis, arthritis is induced in the metatarsophalangeal joint. The patient takes a long time to recover from an attack of gout. In general, the gout attack occurs at night and dawn, when body temperature falls, and continues for about 2 weeks. The inflammation may be detected by an elevated erythrocyte sedimentation rate and higher C-reactive protein levels.

Partial Deficiency of PRPP Enzymes Without Adenosine Deaminase

Immune deficiency in some cases is caused by a lack of the enzyme that converts adenosine into inosine.

$$PRPP + Adenine \rightarrow (Adeninephosphoribosyltransferase) \rightarrow AMP + PP_i$$

$$PRPP + Hypoxanthine \rightarrow$$
$$(Hypoxanthine - Guaninephosphoribosyltransferase) \rightarrow IMP + PP_i$$

$$PRPP + Guanine \rightarrow$$
$$(Hypoxanthine - Guaninephosphoribosyltransferase) \rightarrow GMP + PP_i$$

Gene therapy that returns a retroviral vector incorporating a normal adenosine deaminase (ADA) gene to the patient has been attempted.

PROTEIN BIOSYNTHESIS

Protein biosynthesis is the process by which a cell makes protein. Although this may refer only to translation in a narrow sense, it usually refers to the multistage process of protein biosynthesis from transcription to translation.

Transcription is a process to synthesize messenger ribonucleic acid (mRNA) for protein from genomic DNA. This is done using the nucleotide sequence of one strand called a sense strand or coding strand of the DNA double strands. Transcription consists of the following three steps. RNA polymerase is an enzyme which synthesizes RNA from DNA and binds with the specific sequence on DNA that directs a transcription initiation point. This sequence is called a promoter, and a DNA chain will separate if RNA polymerase binds with a promoter. RNA polymerase moves along with an antisense or template strand, synthesizing a ribonucleotide chain. A sense strand is used for making not another sense strand, but the complementary sequence. Then, the complement or antisense strand of DNA is used for synthesizing the RNA. If polymerase comes to the end, modification is performed in order to enhance the intracellular movement of mRNA. To protect mRNA from degradation, a cap structure and poly(A) tail are added to the 5' and 3' ends, respectively. In eukaryocytes, splicing is performed by an enzyme called a spliceome. The unnecessary intron is removed by this step and the exon is joined to make mature mRNA. Then, the mature mRNA produced is moved into the cytoplasm through a pore in the nuclear membrane.

Start

Before protein synthesis is started, the ribosome dissociates into two subunits of different size. The protein synthesis is initiated by forming an initiation complex (30S initiation complex) assembling a complex of five species of proteins: small subunit (30S), mRNA molecule, formyl methyonyl–transfer RNA, an initiating factor and guanosine triphosphate (GTP). It is aligned to the RNA by assembling the sequence AGGAGGU near the AUG codon. This binding position of the 30S subunit to mRNA is decided by formation of a base pair between the specific sequence AGGAGGU (ribosome joint position) close to an initiating codon AUG and the complementary sequence near the 3' end of the 16S ribosomal RNA (rRNA) molecule. After formation of the 30S initiation complex, a large subunit (50S subunit) binds with this complex, forming a 70S initiation complex. There are two transfer RNA (tRNA) binding sites in a 50S subunit, called the P (peptidyl) and A (aminoacyl) sites.

Extension

When N-formylmethionine (fMet)–tRNA binds with the P site, a tRNA molecule with an anticodon that associates with the codon of the neighboring initiating codon can combine with the A site of the 70S initiation complex.

When acylated tRNA binds with the A site with elongation factor, a peptide bond is formed between the amino acid combined with fMet and tRNA with the peptidyltransferase enzyme complex. In that case, fMet is separated from fMet-tRNA of the P site. In the ribosome, tRNA in the P site that dissociated with an amino acid is released. Next, 2-peptidyl-tRNA moves to the P site from the A site. mRNA moves along these three bases, and the resulting codon appears in the A site. This movement of three is called translocation.

Stop

If extension of a peptide chain proceeds and a codon (UAA, UAG, UGA) is reached from beginning to end, then no acylated tRNA can bind with the A site, and extension of the peptide chain is terminated. Furthermore, the open factor that recognizes these stop signals acts, and the completed peptide chain leaves the ribosome. The ribosome complex dissociates again into the two subunits to be reused for new protein synthesis (Figure 4.20). After

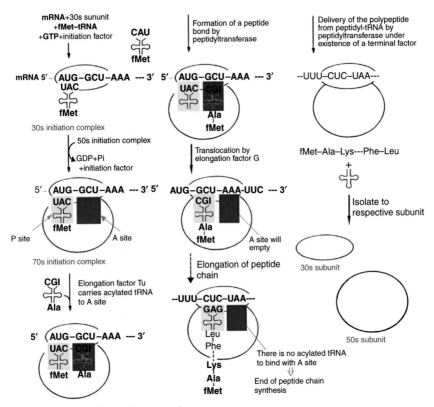

Figure 4.20 *Protein synthesis on ribosomes.*

the completed peptide chain leaves the ribosome, a formyl group (–CHO) is removed by a specific enzyme and, in many cases, the methionine at an end of the N-terminal residue is also removed. Many pieces of ribosomes combine with one molecule of mRNA to efficiently synthesize proteins.

The ribosome complex bound with mRNA is called a polysome, and a polypeptide of 100 residues can be synthesized within 5 seconds.

The acylated tRNA molecule that has an anticodon corresponding to the other codon binds to the A site and the peptide chain is elongated. Modification and folding of the newly synthesized protein will take place cotranslationally. With regard to the folding of the polypeptide chain, in a complete conformation of the protein, called the secondary structure and tertiary structure of protein, most of the protein can be arranged post-translationally. The addition of functional groups, such as formation of a di-sulfide bond, acetyl residue, phosphate residue, lipids and sugar moiety, also occurs post-translationally, resulting in the active conformation of certain proteins. Moreover, the hydrophobic polypeptide chain (signal peptide), which is located at the N-terminal amino acid sequence, is cleaved off by signal peptidase.

REFERENCES

International Human Genome Sequencing Consortium. Initial sequencing and analysis of the human genome. Nature 2000;409:850–921.

Kagawa Y, Nozawa Y, editors. Illustrated Medical Chemistry. 4th ed. Tokyo: Nanzando Co; 2001.

Kunishima C, Inoue I, Oikawa T, Nakajima H, Komoda T, Katayama S. Activating effect of benzbromarone, a uricosuric drug, on peroxisome proliferator-activated receptors. PPAR Res 2007;2007:36092.

Liu H-Y, Liao P-C, Chuang K-T, Kao M-C. Mitochondrial targeting of human NADH dehydrogenase (ubiquinone) flavoprotein 2 (NDUFV2) and its association with early-onset hypertrophic cardiomyopathy and encephalopathy. J Biomed Sci 2011;18:29.

Matsunaga T, Hara A, Komoda T. Oxidized high-density lipoprote. In: friend or foe, in: Komoda T, editors. The HDL Handbook. 2nd ed. New York: Academic Press; 2013. pp. 247–72.

Seo M, Inoue I, Tanaka M, Matsuda N, Nakano T, Awata T, et al. Clostridium butyricum MIYAIRI 588 improves high-fat diet-induced non-alcoholic fatty liver disease in rats. Dig Dis Sci 2013;58:3534–44.

CHAPTER 5

Biochemistry of Internal Organs

Contents

Abstract

This chapter describes the functions of human organs. Breathing controls gas exchange in the lungs and hemoglobin in the red blood cells. Full expansion of the lungs is induced by pulmonary surfactants produced in pneumocytes. Carbon dioxide contributes to the maintenance of physiological pH in the lungs and bloodstream. Carbonic anhydrase is important in the regulation of humoral pH in the aortae, red blood cells and pneumocytes. Waste material in the blood is processed by the kidneys and excreted from the glomeruli as urine. Biological materials and water, calcium (Ca^{2+}) and bicarbonate (HCO_3^-) ions are reabsorbed in the renal tubules. Renal alkaline phosphatases may participate in the reabsorption of Ca^{2+} and HCO_3^- and the maintenance of pH by $H^+ + HCO_3^-$. The gut secretes hormones that control the physiology of the stomach, small intestine, pancreas and gallbladder. The molecular mechanisms of muscle contraction are controlled by relationships between Ca^{2+} and contractile proteins.

RESPIRATORY SYSTEM

Pulmonary Respiration

The function of the respiratory system is to exchange two gases, namely oxygen and carbon dioxide. The gas exchange takes place between several

Biochemistry for Medical Professionals. http://dx.doi.org/10.1016/B978-0-12-801918-4.00005-0

hundred million air sacs (alveoli) in the lungs and the capillary vessels that surround them. Inhaled oxygen moves into the capillary vessels from the alveoli, and carbon dioxide moves to the alveoli from the capillary vessels. In order to take oxygen from the air into the blood that flows through the lungs, three processes are necessary: breathing, diffusion and perfusion. Breathing is the process by which air reaches the lungs. Diffusion is the gas exchange that takes place between the alveoli and capillary vessels of the lungs, without energy consumption. Perfusion is the process by which the blood is sent throughout the lung in the cardiovascular system.

Blood circulation is vital for life, to connect the air that contains oxygen with the cells in the body that consume oxygen. To distribute oxygen to the muscle cells of the entire body, not only the lungs but also the cardio-vascular system, which circulates the oxygenated blood, are important. The respiratory control center is located in the brainstem and breathing occurs spontaneously. It is controlled subconsciously and performed throughout life, from the moment of birth until death.

Protective Mechanisms of the Lungs

A moderately active human breathes about 20,000 liters of air every day. The weight of the inhaled air is 20 kg or more and some poisonous particles and gases are contained in the air. Particles such as dust, smoke, fungal spores, bacteria and viruses can adhere to the respiratory tract or the surface of the alveoli. The respiratory system has protective mechanisms to prevent this invasion. Small hair-like projections called cilia line the surface of the respiratory tract. The cilia beat 1000 times or more every minute. They move the mucous layer that covers the inside of the trachea at a rate of about 0.5–1 cm per minute. The toxic particles and pathogenic pollutants that are caught by this mucous layer are returned to the mouth and coughed out or swallowed into the esophagus.

Owing to the need for gas exchange, the alveoli are not protected by a mucous layer or cilia, since a thick mucous layer would prevent the movement of oxygen and carbon dioxide. Instead, the alveoli have another protective mechanism, namely phagocytes. These mobile cells on the alveolar surface search for foreign particles, combine with them, ingest and, if necessary, kill them, and then they are digested.

The phagocytic cells in the lung are called alveolar macrophages. If these cells are exposed to a threat that presents a serious danger to the lung, they will signal and gather circulating white cells such as neutrophils to eliminate the invaders. Therefore, when a person inhales a lot of dust or has a respiratory tract infection, many macrophages are found in the body and neutrophils gather around the alveoli. This function decreases gradually with aging.

Changes in the lung accompanying aging include not only a reduction in the protective mechanism of the lungs, but also decreases in maximum breathing quantity, the amount of gas exchange and lung capacity. In non-smokers, pathological changes in the physical condition of the lungs rarely occur. However, current and ex-smokers and people with pulmonary emphysema experience more severe breathlessness with age. As a person ages, the muscles of the respiratory systems, such as the diaphragm, are weakened. In healthy elderly people, this decline in muscular power seldom causes problems. However, it can be a serious problem for elderly people with bacterial pneumonia, because the muscular power required for breathing declines further with pneumonia. The muscles of the respiratory system, such as skeletal muscles, can be trained even in elderly people. It is possible to improve breathing by taking daily exercise, including exercises that strengthen the muscles of the respiratory system.

The structural and functional changes in the lung accompanying aging have various effects in elderly people, particularly those who engage in running, cycling, mountain climbing and other strenuous exercise. It becomes difficult to perform intense aerobic exercise. Furthermore, the danger of developing pneumonia through a bacterial or viral infection is increased. The most serious influence is an aging-elicited exacerbation of earlier heart and lung disease; in particular, diseases caused by the adverse actions of smoking become more severe.

Bronchoalveolar lavage (BAL) fluid in the healthy lung has a high surface tension because the lavage contains a high level of dipalmitoyl phosphatidyl-choline (DPPC). Therefore, BAL can be used clinically for the collection of lung surfactant and pathogens. The surface tension of the wash liquid collected from patients with tuberculosis, pulmonary emphysema and bronchiectasis may be low despite normal composition of the phospholipid. In pulmonary fibrosis, the phospholipid composition involved in surface tension is low (approximately 25%), suggesting a qualitative alteration in lung surfactant based on the lipid metabolism abnormality. BAL collection does not require a thoracotomy and repetitive implementation is possible; therefore, density gradient ultracentrifugation may be the best method for the quantitative analysis of lung surfactant, including surfactant protein A (SP-A), SP-B, SP-C, SP-D and alkaline phosphatase. Proteins and other constituents of lung surfactant can be measured using the BAL analysis data, which may be useful in the elucidation of non-respiratory pulmonary function or lung cancer.

Indeed, experiments on lung cancer specimens have shown that there is a lack of high-density surfactant in lung cancer (Iino et al., 2007). The lungs are active organs that are involved in several functions, not only respiration but also metabolism and immunity. Administration of preventive surfactant medication to embryos resulted in a decreased risk of respiratory distress syndrome (RDS). In other embryos, surfactant medication originating from porcine lung reduced the rate of development of symptoms of pneumothorax and interstitial pulmonary emphysema (Choi et al., 2005). However, a standard needs to be established to identify those embryos at risk of RDS who may benefit from this treatment. A surfactant-like particle (SLP) is located in the small intestine, and antihuman SLP antibodies cross-react with human pulmonary surfactant. Therefore, human SLPs could be useful for evaluating the function of pulmonary surfactant in metabolic syndrome. Metabolic syndrome can be induced in knockout mice in which intestinal alkaline phosphatase (IAP) has been inactivated; this syndrome can then be treated with bovine IAP (Kaliannan et al., 2013). Recombinant IAP has been used for this treatment in many animals. Seo et al. (2013) report that the medicine MIYAIRI 588, extracted from *Clostridium butyricum*, could be useful for the treatment of non-alcoholic fatty liver disease in rats with metabolic syndrome.

Change in Oxygen Partial Pressure

While atmospheric air has a pressure of 760 mmHg, the partial pressure of oxygen (pO_2) is approximately 160 mmHg in atmospheric air, 102 mmHg in alveolar air, 100 mmHg in arterial blood and 40 mmHg in venous blood.

The diffusion of oxygen is determined by the pressure gradient in the alveoli, the permeability of the boundary membranes and the area available for diffusion. The pressure gradient of an alveolus and mixed venous blood is 60 mmHg. The thickness of a boundary membrane alveolar wall is 0.5 μm or less and the total surface area in an adult human is as great as 60–70 m^2. Although it takes about 0.25 seconds for the formation of oxygen balance between the alveoli and the blood, the time taken to flow through a lung capillary vessel is 0.75 seconds at rest.

Oxygen Dissociation Curve

The oxygen dissociation curve is a graph with oxygen partial pressure along the horizontal axis and oxygen saturation on the vertical axis, which shows an S-shaped relationship. Oxygen and carbon dioxide are transported in

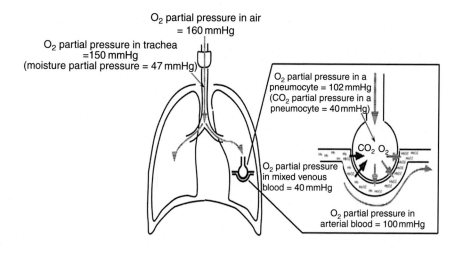

O_2 partial pressure in air
= 160 mmHg

O_2 partial pressure in trachea
=150 mmHg
(moisture partial pressure = 47 mmHg)

O_2 partial pressure in a
pneumocyte = 102 mmHg
(CO_2 partial pressure in a
pneumocyte = 40 mmHg)

CO_2 O_2

O_2 partial pressure
in mixed venous
blood = 40 mmHg

O_2 partial pressure in
arterial blood = 100 mmHg

Figure 5.1 *Transport of oxygen (O₂) and carbon dioxide (CO₂) in the lung and bloodstream.*

the blood as a result of changes in blood partial pressures (Figure 5.1). Most oxygen is taken into the hemoglobin in red blood cells, although trace levels of oxygen exist in the dissolved form. The oxygen content is about 20 ml, when is hemoglobin at 15 g per 100 ml of blood is 100% saturated with oxygen. It has the characteristic advantages of taking in oxygen via the lungs and dissociation of oxygen in organs. With increased carbon dioxide excretion, increased hydrogen ion (proton, H^+) concentration (fall in pH) and increased partial temperature, the oxygen dissociation curve is shifted to the right, promoting oxygen dissociation. At this time, the affinity of hemoglobin for oxygen (P_{50}) becomes large.

Oxygen Consumption

Twenty five percent of oxygen is contained in arterial blood at any one time. Only 5 ml of oxygen per 100 ml of blood is consumed, and 15 ml of oxygen is returned without being used. If the cardiac output in 1 minute is about 5000 ml, the oxygen consumption will be 250 ml. Carbon dioxide produced by metabolism reacts with water and is converted to carbonic acid (H_2CO_3), which dissociates into hydrogen (H^+) and bicarbonate (HCO_3^-) ions:

$$CO_2 + H_2O \Leftrightarrow H_2CO_3 \Leftrightarrow H^+ + HCO_3$$

Transport of Carbon Dioxide by Red Blood Cells

Although the above reaction takes several seconds in tissue fluid or plasma, since red blood cells contain carbonic acid anhydrase, an enzyme that accelerates the reaction, carbon dioxide can be processed in several milliseconds. The red blood cells excrete carbon dioxide. Some of the carbon dioxide is bound with hemoglobin (Hb). It may be carried as + Hb-NH$_2$-Hb-NHCOOH, but most is carried in the form of bicarbonate ions (HCO$_3^-$). Glycosylated hemoglobin (HbA$_{1c}$) is produced in diabetes and can be measured as a marker of this disease. The concentration of HbA$_{1c}$ is low in young people and high in elderly people. This phenomenon may depend on the amount of reactive oxygen species.

Carbon Dioxide Dissociation Curve

If the partial pressure of carbon dioxide is plotted along the horizontal axis and carbon dioxide content on the vertical axis, the relation will be a straight line. This carbon dioxide dissociation curve will be shifted upwards if the degree of oxygen saturation falls, and will move in the direction in which the carbon dioxide content increases. That is, the content of carbon dioxide in the venous blood increases more than in arterial blood. The partial pressure of carbon dioxide is 46 mmHg in mixed venous blood and 40 mmHg in arterial blood. Since carbon dioxide is extracted from 100 ml of blood under resting conditions, if cardiac output is 5000 ml/minute, the rate of carbon dioxide excretion will be 200 ml/minute. Most of the carbon dioxide produced by metabolism is stored in plasma as bicarbonate (HCO$_3^-$) after gas exchange in the lungs. These bicarbonate ions play an important role in buffering the pH of blood. For example, the reaction lactic acid {ReversReact} H$^+$ + lactate ion$^-$ takes place, sodium lactate is formed, the acidic H$^+$ is taken into a carbonate (H$_2$CO$_3$), and the reaction is H$_2$CO$_3$ {ReversReact} CO$_2$ + H$_2$O. Carbon dioxide from this reaction is excreted as gas from the lungs. Carbon dioxide dilates the blood vessels and thus increases blood flow.

RENAL AND UROLOGICAL SYSTEM

Filtration in Glomeruli

Urea is filtered from the blood in the glomeruli of the kidneys, and the urine volume produced per day is 100 liters. The nephric tubules reabsorb 99% of the urine in the renal tubule and the remainder is excreted as urine. The creatinine clearance value (ml/minute) is clinically tested as a marker for checking the function of glomerular filtration. The amount of creatinine

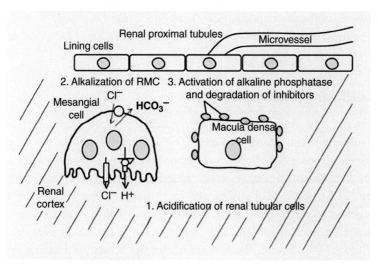

excreted is affected by meals and the circadian rhythm. Because creatinine is formed from creatine by catalysis of creatine phosphokinase, the amount of creatine is also regulated by the circadian rhythm.

Reabsorption in Renal Tubules

The renal tubules collect some urinary components after glomerular filtration, and the recovered biological components can be reused in the metabolic systems in the body.

Secretion in Renal Tubules

The secretion of IAP and glycosylphosphatidylinositol (GPI) anchor proteins such as Tamm–Horsfall protein in the urine may be reused in the distal convoluted tubule or in the bloodstream, and can control the host-defense mechanism from lipopolysaccharide molecules (Vyletal et al., 2010). Therefore, urinalysis is a useful marker of infection.

Regulation of Humoral pH in the Kidney

Homeostasis of blood pH is maintained by the action of carbonic anhydrase in the kidney (Kaunitz and Akiba, 2006). The maintenance of pH, and cotransportation of calcium and phosphorus ions, may involve carbonic anhydrase and alkaline phosphatase in the kidney under the influence of aldosterone (Figure 5.2).

Figure 5.2 *Possible mechanism of kidney alkaline phosphatase related to acid-base balance with carbonate/bicarbonate.*

Hormonal Disorders

Urological disorders in older women may originate in the decrease in gonadal hormones (e.g. estrogen) linked with aging.

In contrast to female hormones, secretion of male hormones (androgens, e.g. testosterone) decreases slowly with age, leading to a syndrome called partial androgen decline in the aging male (PADAM) or andropause. Testosterone can be injected intramuscularly in men with PADAM. This treatment is contraindicated in aging patients with prostate carcinoma, urinary disorders or benign prostatic hyperplasia.

GASTROINTESTINAL ORGANS

The human digestive system includes the internal organs from the mouth to the anus. The human gut hormones and their related enzymes are located in the stomach, pancreas, gallbladder and small intestine.

Gastrin promotes the secretion of gastric acid (hydrochloric acid, HCl). Pepsin secretion is activated by HCl in the stomach.

Enterochromaffin-like cells secrete histamine. Substance P promotes peristalsis, and vasoactive intestinal peptide (VIP) promotes vasoconstriction and peristalsis.

Cholecystokinin is secreted from I cells in the small intestine. It promotes the secretion of various peptidases, amylases and lipases from the pancreas, and bile secretion from the gallbladder.

Gastric inhibitory peptide (GIP) is an endocrine hormone.

Secretin is secreted from S cells in the duodenum, and H_2CO_3 secretion to the intestine can promote uptake from the pancreas.

MUSCULAR SYSTEM

Movement of the skeletal muscle system results from conformational changes in the motor proteins myosin and actin. The fine structure model was reported from observations made by electron microscopy of the complex of actin and myosin. Myosin contains six subunits, including two long entwined fibers or tails attached to large head groups, and four smaller chains. The myosin fiber is comprised of a light chain (light meromyosin or LMM) and a heavy chain (heavy meromyosin or HMM). The head attaches to the actin molecules.

When ATP is bound, the myosin undergoes structural changes which trigger the breakdown of adenosine triphosphate (ATP) to adenosine diphosphate (ADP) by the myosin. The phosphate is released in a mechanism

Muscular tissue
(muscular filament)

Muscle fibril

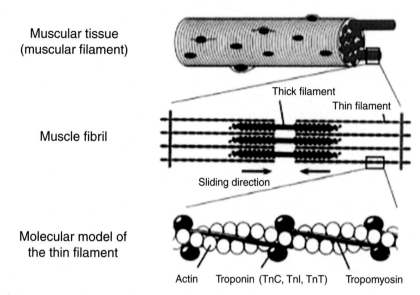

Thick filament

Thin filament

Sliding direction

Molecular model of
the thin filament

Actin   Troponin (TnC, TnI, TnT)   Tropomyosin

Figure 5.3 *Molecular structure of skeletal muscle.*

in which the myosin binds, moves in an action called the power stroke, and releases, resulting in movement within the sarcomere. Skeletal muscle myosin II was one of the first to be discovered, with muscle myosin being only one protein of a large genetic superfamily. Troponin proteins are essential for the process of myosin binding to actin. Figure 5.3 shows the arrangement of tropomyosin and actin within the muscle fibril.

REFERENCES

Choi CW, Hwang JH, Yoo EJ, Kim KA, Koh SY, Lee YK, et al. Comparison of clinical efficacy of Newfactan versus Surfacten for the treatment of respiratory distress syndrome in the newborn infants. J Korean Med Sci 2005;20:591–7.

Iino N, Matsunaga T, Harada T, Igarashi S, Koyama I, Komoda T. Comparative characterization of pulmonary surfactant aggregates and alkaline phosphatase isozymes in human lung carcinoma tissue. Cell Tissue Res 2007;328:355–63.

Kaliannan K, Hamarneh SR, Economopoulos KP, Alam SN, Moaven O, Patel P, et al. Intestinal alkaline phosphatase prevents metabolic syndrome in mice. Proc Natl Acad Sci USA 2013;110:7003–8.

Kaunitz JD, Akiba Y. Duodenal bicarbonate – mucosal protection, luminal chemosensing and acid-base balance. Aliment Pharmacol Ther 2006;24(Suppl 4):169–76.

Seo M, Inoue I, Tanaka M, Matsuda N, Nakano T, Awata T, et al. Clostridium butyricum MIYAIRI 588 improves high-fat diet-induced non-alcoholic fatty liver disease in rats. Dig Dis Sci 2013;58:3534–44.

Vyletal P, Bleyer AJ, Kmoch S. Uromodulin biology and pathophysiology – an update. Kidney Blood Press Res 2010;33:456–75.

CHAPTER 6

Biotechnological Study

Contents

Abstract

This chapter outlines practical technology and experimental conditions using up-to-date biochemical technology. Regarding high-performance liquid chromatography (HPLC) and HPLC–mass spectrometry (MS), computer-linked MS or MS-MS can identify substances from their molecular weight. The fluorescence-activated cell sorting method can identify species of cells immunohistochemically. Western and Northern blots are described here and can be replicated by readers of this book. Polymerase chain reaction (PCR) techniques can be performed using commercially available kits that may be easier to use than real-time PCR. The key point is how to design primers for certain messenger RNA. The creation of knockout or knockdown mice can suppress the target gene by genetic manipulation, or the target gene can be suppressed by interfering RNA. Induced pluripotent stem cells can be generated from adult cells and have great potential in medicine.

Biochemistry for Medical Professionals. http://dx.doi.org/10.1016/B978-0-12-801918-4.00006-2

HIGH-PERFORMANCE LIQUID CHROMATOGRAPHY AND HIGH-PERFORMANCE LIQUID CHROMATOGRAPHY–MASS SPECTROMETRY

Components of High-Performance Liquid Chromatography

- Solvent
- Solvent change valve
- Pump
- Pressure damper
- Mixer
- Rheodyne® manual injector
- Column
- HPLC controller
- Detectors (e.g. ultraviolet absorbance meter)
- Detection data analogue/digital (A/D) conversion equipment (data output equipment)
- Personal computer (for data logging and data analysis)
- Printer

The pump is the core of high-performance liquid chromatography (HPLC). It has a structure that is extremely stable for fluid transmission. An online deaeration unit is attached in many cases. The upper limit of the pump pressure in HPLC is 40 MPa, whereas an ultra–performance liquid chromatography (UPLC) system can reach a maximum of 100 MPa; this technology is relatively new and details are not given here.

Single-plunger pumps were the standard specification in the classic system, and frequently used for sending the fluid of the main mobile phase. Because of advantages such as ease of use and maintenance of inexpensive apparatus in which the analysis does not require high accuracy, this specification is now commonly used for fluorescent labeling reagents. However, to reduce the mobile phase pulsation observed in the single-plunger pump, gradient analysis is becoming more common. In the gradient system, there are two types of dual-pump system: co-current and direct current. Pumps may be classified according to the analysis method used. An isocratic pump can send fluid using only one kind of moving bed. The pump is not expensive, but since gradient analysis cannot be performed the analytical width is narrow. Two or more sets of pumps are connected with a high-pressure gradient pump, each flow velocity is controlled and gradient analysis is conducted. Since it is necessary to prepare a pump for every kind of moving bed, this is expensive. Although a low-pressure gradient pump has only one pump, the valve that sets four solvents in motion is attached to the front of the pump and conducts gradient analysis. The pump pressurizes after mixing the solvent. It is less expensive than a high-pressure gradient pump and solvent mixing is easy. A nanoflow or microflow pump sends fluid in a flow (nanoliters or microliters) suitable for analysis of samples of ultralow volume. Usually, a capillary column is connected. An analysis pump ejects fluid at a flow rate suitable for normal analysis (about 0.1–10 ml/minute). For separation refining of a preparative isolation pump substance, fluid is sent at a flow rate of 10 ml/minute or more. This kind of pump often has a lower resistance pressure than the object for analysis.

The sample is poured in via an injector, which may be a manual or an automatic type. Since most injectors are produced by Rheodyne, manual injectors are commonly known as rheodynes. The set-up can be easily changed using two kinds of channel. First, a port is made on the loading side and a syringe is inserted in a sample loop. The sample should be accurately inserted using a syringe and a sample loop equipped in the injector. Although the accuracy depends on the skill of the operator, the use of a sample loop is in general more quantitative than that of the syringe because any excess sample overflowing from the sample loop is drained. Second, the port is moved to the injection side and the injected sample is poured into the flow to the column.

If a function that outputs an electrical signal to a manual injector is attached, an injection signal can be sent to a detector or an integrator at this time. Almost all HPLC equipment uses the Rheodyne manual injector and measures the sample by syringe, adjusts it and pours it into the channel. The manual method of sampling is carried out according to the manufacturer's instructions, and can be selected according to its purpose. A sample is put into a plate or two or more vials, and placed in the equipment for serial analysis. Another part of the equipment maintains the correct temperature. The mixture in the sample is separated by column chromatography. The column is made from stainless steel, filled with ball-shaped porous silica gel with an alkyl residue. For preparative isolation, pulverized silica gel is used. As the silica gel has a smaller diameter, a higher resolution pattern of peaks is obtained, although this requires a high pressure to pump the fluid. Therefore, reduction in the pump pressure is accomplished by various devices, which include an increase in the resistance pressure of the flow passing between pump and injector or between injector and column, and column heating to lower the solvent viscosity. Since the difference in the various kinds of HPLC is often dependent on the column type used, it may be possible to perform and optimize the process without equipment modification by changing only the column. However, since the column may be damaged by using different types of solvent, caution is needed when changing solvents.

A column heater or column oven surrounds a column and performs heating or cooling to the temperature of the column. When the temperature of the column circumference is unstable, the elution time may not be stabilized and reproducibility may be worsened. Therefore, a heater is attached to keep the temperature constant. The column temperature is used as one of the isolation parameters. Although it is called an oven or a heater, the equipment also has a cooling function, which can be useful for analysis at specific temperatures.

Chromatography Detectors

Different kinds of detector may be used according to the characteristics (optical properties, electrochemical properties and mass spectrometry) of the substance being analyzed. General rules on HPLC are published by organizations such as the Japanese Standards Association (JIS K0124:2002).

The detector equipment includes the spectroluminous-intensity detector (UV/VIS detector), fluorescence detector (FLD), differential refractometer (RID), electrochemical detector (ECD), conductometric detector (CD), mass spectrometer (MS), infrared spectrophotometer (IR), angle-of-rotation detector (OR), circular diachronic detector (CD), flame ionization detector (FID), radioisotope detector (RI), dielectric constant detector, chemiluminescence detector (CLD), atomic absorption analysis equipment (AA), inductively coupled plasma emission–spectrochemical analysis equipment (ICP-AES), high-frequency plasma mass spectrometer, heat detector, optical distraction detector, viscosity detector, ion electrode, ultrasonic detector and nuclear magnetic resonance (NMR).

Absorbance detectors use the characteristic of absorbance of light of a specific wavelength, i.e. ultraviolet and visible light. These detectors include diode array detectors, which can simultaneously detect a broad wavelength. The three dimensions of time, absorbance and wavelength can also be displayed as a result. Angle-of-rotation detectors (optical rotatory detectors) are used for measurement of rotatory polarization or optical activity. Circular dichroism detectors use polarization properties and are used for the detection of optical isomers.

Characteristics of fluorescence are detected using fluorescence detectors. A beam of light is used to excite the substance, which then emits light. For a substance that does not fluoresce, detection is possible by labeling or tagging the substance.

Refractive index detectors use the angle at which light passes through a sample as its concentration changes. These detectors cannot be used in gradient analysis. They have low sensitivity but can be used on most substances.

Evaporative light scattering detectors detect a ray scattered from a compound by radiating light. Before detection of the scattering ray, the compound is isolated by evaporating the mobile phase. Therefore, the detector can be used for detection of only non-volatile compounds. Despite initial difficulties and few improvements in sensitivity and operativity, their performance has improved with growing requirements in the field of medicine. Since the gradient eluting method is used in detection, it is effective in the distributing method of sugars, especially oligosaccharides, because of 5–10-fold increases in the signal-to-noise ratio compared with a differential refractometer.

Electrochemical detectors measure changes in pH or oxidation–reduction potential. The detectors monitor electrical conductivity from conductometric substances.

Electrospray Ionization

In electrospray ionization (ESI), atmospheric pressure chemical ionization (APCI) or matrix-assisted laser desorption/ionization (MALDI) can be used as the ion source. Detectors include a primary detecting element, quadruple type ion trap and time of flight (TOF) ion trap (IT). The sample labeled with the radioactive isotope is measured. The electrical signal output from the detector is recorded and is interpreted by detecting its peak. Results are printed on thermal paper. If the equipment is not controlled, and if it uses a detector, it will be satisfactory; but to control equipment simultaneously, the same manufacturer's apparatus should be used. A Windows PC is often used as an integrator. In other types of equipment, the exclusive design is carried out using integrated operating systems. When not controlling equipment, normal-phase chromatography was first used in HPLC using the manufacturer's apparatus.

High-polarity silica gel is used for the stationary phase, and low-polarity material for the mobile phase. The isolation of the substance under analysis is carried out in the stationary phase so that the polarity is higher and elution is slower. In a mobile phase with many rates, a substance with high polarity will be eluted sooner. Although the use of normal-phase chromatography has decreased in recent years with the development of reverse-phase chromatography, since the normal-phase type has different characteristics from other separation modes, including the reverse-phase type, it can be very effective depending on the purpose. For example, the mobile phase can be used to conduct analysis of tocopherol isomers that are difficult to separate and sugars that are difficult to retain by reverse-phase chromatography. It does not usually contain water. It is suitable for the separation of compounds, such as acid anhydrides, which can be easily hydrolyzed. Reverse-phase chromatography is a variation of normal-phase chromatography. It uses not only the silica gel stationary phase of conventional normal-phase chromatography but also the stationary phase that introduces the surface of polar compounds. Therefore, chemicals are separable by the mobile phase containing water that could not be used in a silica gel stationary phase. The normal-phase mode in reverse-phase chromatography can be used to separate polar compounds, such as complex amino acids. In reverse-phase chromatography, a substance of low polarity is used for the stationary phase, and organic solvents, such as a solution of salts of high polarity, alcohol and acetonitrile, are used for the mobile phase.

When the mobile phase pH for separation needs to be set outside the range of use for silica gel, or when the unreacted silanol residue which remains on the silica gel surface adversely affects separation and this cannot be solved by changing the mobile phase, resin may be used as a stationary phase. Analysis of the sample is carried out on the stationary phase so that polarity is lower and elution is slower. Moreover, elution using a mobile phase for many types of substance with low polarity occurs sooner. Octadecyl silica, which consists of an octadecyl residue with 18 carbon chains with silica gel, is used extensively in ODSc columns. Reverse-phase chromatography has been used for the analysis of low molecular weight substances, and is now also used for nucleic acid and protein analysis. Proteins are analyzed by the gradient eluting method, usually with pH 2–3 or near neutral pH as mobile phase conditions, with an incremental rate of organic solvents, using chemically modified silica gel with large pores as the column filler (Wilson et al., 2005).

FLUORESCENCE-ACTIVATED CELL SORTING

Fluorescence-activated cell sorting (FACS) measures the antigen levels on the cell surface quantitatively. Cells are dyed with a fluorescent antibody, then placed in a stream of liquid which passes the focus of a laser, and each cell emits light. Not only the fluorescence intensity of the inside and outside of a cell but also its biological characteristics, such as size (by forward-scattered light) and internal structure (by side-scattered light), can be measured. FACS can be divided into apparatus in which only the analysis of each parameter is possible, and cell-sorting apparatus that has a preparative isolation function, which can isolate a specific cell preparatively and abacterially (sorting). There are drop charge systems and cell-capture systems in the sorting mechanism (Tomlinson et al., 2012).

WESTERN BLOT

The wet protein gel and a nitrocellulose membrane are sandwiched with filter papers wetted with buffer solution. The power is set up and applied to the negative polar plate of the translational apparatus. The positive polar plate of the top board is piled with filter papers before the transfer. By application of an electrical field, protein transfer from the gel to the nitrocellulose is completed in 90 minutes. The specific antibody for the protein under investigation is added to the nitrocellulose. An antibody against the target protein is fixed specifically on the nitrocellulose. After washing out the excessive antibody, the protein–antibody adducts on the nitrocellulose membrane are detected with the secondary antibody, specific to the animal species of the first antibody. It can be detected by the following two methods.

1. Using the secondary antibody by which the signal was found affixed to an enzyme called horseradish peroxidase (HRP). Detection is by the fluorescence reagent that emits light as a substrate of HRP. If this membrane is exposed to X-ray film, a membrane is exposed by the fluorescence from the band of the target protein, and will be detected as a black line (Figure 6.1).

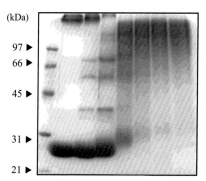

Figure 6.1 *Western blotting of various high-density lipoproteins. Source: Modified from Matsunaga et al. In: The HDL Handbook (2010, p. 220).*

2. Using a secondary antibody to which a fluorescent substance has bound. Since the band will show fluorescence if the excitation light is hit, this can be read by specific detectors.

NORTHERN BLOT

Northern blotting is the standard method for investigating quantities such as mRNA size. The usual method uses a DNA probe after RNA denatured by glyoxal has been transferred to the membrane following electrophoresis in agarose gel, labeled with $[^{32}P]$deoxycytidine triphosphate for radioactive detection, or a biotinylated probe detected by chemiluminescence.

The procedure is as follows.

1. Gel electrophoresis of extracted RNA is carried out, using high-quality RNAse-free reagents and a comb for production of wells in the gel. Electrophoresis is carried out at 60 mA for about 10 minutes, then at 50 mA for 2–3 hours.
2. After running the gel, the RNA is detected by staining. Put the gel into a tray for 20 minutes or more with 12.5 μg/ml ethidium bromide (hazardous) at room temperature and swirl gently. Take a photo on a transilluminator. On the photo, examine the size of the bands relative to the position of 28S and 18S.
3. Put the gel in the blotting tray. Shake gently for 15 minutes with 10 × saline–sodium citrate (SSC) buffer.
4. The RNA is transferred to either a nylon or nitrocellulose membrane. For Glyoxal release and membrane processing using GeneScreen Plus®, mark the position of the upper well with a ballpoint pen. Follow the manufacturer's directions for the washing of the gel and transfer to the membrane.
5. Prehybridization/hybridization: The prehybridization time is longer than usual. At 55–65°C, run prehybridization for 1 hour, retain the membrane and discard the solution after use. Use the recommended prehybridization solution. The hybridization solution described in step 6 is then added. For prehybridization and hybridization, a small tray, plastic bag or other device is typically used.
6. Hybridization solution: Use ssDNA at a final concentration of 300 μg/μl and 5–20 ml of hybridization solution as recommended for the membrane. Make sure that no air is present. Hybridize at 55–65°C in a water bath with gentle shaking for 8 hours.

7. Follow the manufacturer's instructions for washing the membrane. A radioactive probe can be detected in this way: the membrane is wrapped in plastic wrap, exposed to an imaging plate, and detected in a Typhoon 8600 or BAS 2000 or other image detection system. For biotinylated probes, an appropriate chemiluminescent detection system should be used.

POLYMERASE CHAIN REACTION

Relative Quantitation of Messenger RNA Expression Level

When comparing the mRNA expression level of a certain gene using real-time polymerase chain reaction (RT-PCR), the target gene is amplified in real-time PCR and the expression level is calculated using the cycle threshold (Ct) values. For an accurate comparison, RNA must be extracted from the same number of cells, and amplification of the target RNA must be conducted at the same level of efficiency. However, since such conditions will not be fulfilled in reality, it is necessary to adjust the expression level to that of a housekeeping gene. The procedure of relative quantitation is explained using an example.

Fundamental View of Relative Quantitation

Theoretically, since amplification products double in number every cycle, the difference in Ct value implies exponentiation of two, but PCR does not follow this theory. Therefore, the amplification efficiency should be compared for accurate quantification, even if the products are amplified using the same primers. The cycle reflecting this amplification efficiency is an analytical curve. The analytical curves are amplified with a series of samples of differing concentration, making a graph of the concentrations of the samples, and obtaining the relative Ct values. When measuring the relative amounts of samples, they can be compared on the analytical curve. It is necessary to prepare this analytical curve for every target gene and primer used. Differences in the target gene and primer will affect the amplification efficiency. If PCR conditions vary, even if the same primers are used, an analytical curve should be constructed independently. Then, it should not be quantified only about the target gene. Expression of the housekeeping gene should be quantitated as an internal standard in every sample to normalize expression of the target genes. Even if the same amount of RNA is used for the reaction between samples, it will vary for each cell sample from which it originates. The total expression level of

RNA in cells is also different according to the culture conditions. Even if the RNA size of an internal standard is constant, the size of the whole mRNA will differ. Therefore, it alters the quantity of RNA and the initial number of cells will differ. When measuring an expression level, the RNA is relevant only to a particular cell.

Preparing samples with the same number of cells does not guarantee that the extraction efficiency of RNA will be identical. For some of these factors, compensation is needed if the method of relative quantitation is simple packing. For a housekeeping gene with constant expression level, RT-PCR is performed based on the RNA solution of a certain regular quantity. A fixed quantity of the expression level of the housekeeping gene is performed based on an analytical curve from the Ct value, and the relative value based on a certain sample is calculated. This shows how many errors have arisen in the RNA solution used between samples. Next, RT-PCR is performed based on the RNA solution of the quantity of a target gene. The expression level of a target gene is quantified based on an analytical curve from the Ct value. The fixed-quantity result obtained is divided by the relative value, and the error of RNA used for the reaction is corrected. Based on the fixed-quantity result after this adjustment, the relative value of the target gene between samples is calculated.

Conditions of Relative Quantitation

As in the preceding section on the preparation of the sample for creating an analytical curve, a fixed quantity of the expression level of the housekeeping gene and target gene requires creation of an analytical curve. RNA or cDNA is used as the sample for analytical curve creation. These preparative methods are described below.

Preparation of the RNA Sample to Create an Analytical Curve

When analyzing a fixed quantity in RT-PCR, RNA is desirable as the sample for preparing an analytical curve. The reaction efficiency of a reverse transcription process will be disregarded in the case where an analytical curve is prepared with a cDNA sample. If it is carried out with the RNA sample, the analytical curve in consideration of the reaction efficiency of both reverse transcriptase and PCR can be prepared. The following can be used as the RNA sample for analytical curve creation.

A fixed quantity is used and the relative value on the basis of a certain sample is calculated. This shows the number of errors in the RNA solution used between samples. Next, RT-PCR is performed based on the RNA solution of the quantity about the target gene used.

Manufacture of Artificial Composition RNA

1. The RNA arrangement (housekeeping gene) of interest is amplified by RT-PCR from an RNA sample, and double-strand DNA with the sequence of interest is prepared. Next, *in vitro* transcription is performed by making this into an internal standard, and the RNA transcript (single-strand RNA) is prepared.
2. Serial dilution of RNA transcript is carried out and used as the sample for analytical curve creation, after calculating a copy number.
3. A DNA-free check is performed.

Manufacture of the Standard by Real-Time Reverse Transcription–Polymerase Chain Reaction

A one-step RNA-PCR kit is used. In the case of rat tissue non-specific alkaline phosphatase (TNAP), RT-PCR amplification is carried out using 500 ng total RNA or 200 ng mRNA as the template.

- Inner primer 1
- Forward primer CCTCC
- Reverse primer GTCAA
- Inner primer 2
- Forward primer GGCGT
- Reverse primer GCTCC

T7 promoter and oligo-dT (5'-AAGGCCACCAATGCTTTTAT) are added to the forward and reverse primer, respectively. After checking amplification, the primers are purified in a column and prepared at a concentration of about 20–40 ng/l.

In Vitro *Transcription*

In vitro transcription of standard double-strand DNA is performed using a T7 transcription kit. DNase I processing decomposes standard DNA contained in a transcription product. After extraction with phenol chloroform and chloroform/isoamyl alcohol, the RNA fraction is subjected to isopropanol precipitation and then dissolved in RNase-free H_2O. The optical density at 260 nm (OD_{260}) is measured; the concentration is read and a copy number is calculated from the concentration.

Dynamic Range Check

The values of the dynamic range of the RNA samples must be checked. A fixed quantity is used and the relative value on the basis of a certain sample is calculated. This shows the number of errors in the RNA solution used between samples. Next, RT-PCR is performed based on the RNA quantity of a target gene. The expression level of the target gene is quantified based on an analytical curve from the Ct value. The fixed-quantity result obtained is divided by the relative value acquired, and the error of RNA used for the reaction is corrected. Based on the fixed-quantity result after this compensation, the relative value of the target gene between samples is calculated.

Conditions of Relative Quantitation

As in the preceding section on preparation of the sample for creating an analytical curve, a fixed quantity of the expression level of the housekeeping gene and target gene requires creation of an analytical curve. RNA or cDNA is used as this sample for analytical curve creation. These preparative methods are described below.

Preparation of the RNA Sample to Create an Analytical Curve

When using a fixed quantity in RT-PCR as the sample for preparing an analytical curve, RNA is desirable. The reaction efficiency of a reverse transcription process will be disregarded in the case where an analytical curve is prepared with a cDNA sample. If it is carried out with an RNA sample, the analytical curve in consideration of the reaction efficiency of both reverse transcriptase and PCR can be prepared. The following can be used as an RNA sample for analytical curve creation.

A fixed quantity is used and the relative value on the basis of a certain sample is calculated. This shows the number of errors in the RNA solution used between samples. Next, RT-PCR is performed based on the RNA solution of the quantity about the target gene used.

Manufacture of Artificial Composition RNA

1. The RNA arrangement (housekeeping gene) of interest is amplified by RT-PCR from an RNA sample, and double-strand DNA with the sequence of interest is prepared.
2. *In vitro* transcription is performed by making this into an internal standard, and an RNA transcript (single-strand RNA) is prepared.

Implementation of Real-Time Polymerase Chain Reaction

Once the sample for analytical curves and a sample to quantify are ready, the reaction can proceed and the amplification curve can be drawn. In two-step RT-PCR, the primer of reverse transcriptase can be chosen from oligo(dT) primer, random primer and the downstream primers used by PCR. Since some will not have been completely extended by reverse transcription, oligo(dT) primer is seldom suitable for carrying out PCR amplification of domains far from the poly(A) arrangement. Random primer may generate a fluorescence signal in which random primer carried into the PCR anneals with a probe, when using the cycling probe method. In two-step RT-PCR using cDNA for PCR, the difference in reverse transcriptase efficiency can be disregarded if the reaction of a housekeeping gene (internal standard) and a target gene is used with cDNA of the same reaction origin as the sample. Here, the relative quantitation of a human ALP gene is shown as an example. It can also be done in an RT-PCR Core Kit as a reaction reagent.

- Target gene: IAP
- Internal standard: RNA sample for β-actin
- Unknown sample: RNA transcript for three types, A–C (A: drug administration; B: proximal small intestine; C: distal small intestine)
- Correlation sample: IAP, seven steps (10-fold dilution series).

Result of ALP and Calculation of Amplification Result
Relative Quantitation of Sample
Relative quantitation is calculated based on the fixed-quantity result obtained by Smart Cycler. In the IAP experiment, it is calculated as follows.

Computational Procedure
Errors in the amount of RNA are calculated using the fixed-quantity result for β-actin (internal standard). It is normalized to the error calculated in the fixed-quantity result of IAP. The relative value between samples is calculated based on the fixed-quantity result of CYP1A1 compensation. End product: The ratio of the expression level of IAP is calculated as sample $A : B : C = 1 : 33.6 : 1.69 \times 10^4$. If an absolute numerical value is used for analytical curve creation about a fixed quantity and an actual expression level, the result obtained can be a fixed quantity. However, this result is not the same as the true expression level, because it is impossible to extract all the RNA from a sample. Furthermore, since the yield of RNA molecules of the target gene collected is unknown, the true expression level cannot be drawn from the obtained result. Although an absolute fixed quantity of the target gene in extracted RNA is made eventually, the true expression level cannot be calculated. This is because of the problem of whether it adheres to something or can perform relative quantitation between genes. A fixed quantity is absolutely performed for each gene and the result is compared to measure the relative expression level of different genes. This is only one

method. This absolute fixed quantity does not reflect the actual expression level certainly, as stated previously, and it only reflects the quantity inside the extracted RNA. It is necessary to investigate in advance whether it has such a guarantee, although the RNA molecule of every gene should be extracted at the same efficiency. It investigates the range (maximum and minimum) of a quantitation limit, the number of cycles of concentration, and whether dependent amplification can be checked or an actual sample lies within its limits.

RT Reaction
- Used amount of RNA: primer for total RNA 10 ng
- Reverse transcription: oligo(dT) primer
- Reaction capacity: 1–10 μl
- Reaction temperature, time: 42°C, 10 minutes.

PCR
- Used amount of cDNA(s): 10 μl RT reaction mixture/25 μl reaction mixture
- Liquid reaction condition: set up the optimal conditions for each target
- Setup by Smart Cycler software
- Sample type: correlation sample to be standardized and quantified
- Unknown sample: quantitative value of a sample; Standard/Reference: correlation sample.

Plasmid Used for Cloning of Target Polymerase Chain Reaction Product
In this case, a slightly longer PCR product that contains a fragment to amplify or its internal fragment is cloned into plasmid, and this plasmid is used as a sample for analytical curves. The copy number is found as in the preceding section, and serial dilution of the solution is carried out. To use the plasmid as the correlation sample, it is necessary to take care so that the primer does not make a mistake annealing the DNA sequence on the plasmid.

KNOCKOUT AND KNOCKDOWN

When conducting performance analysis of a certain specific gene (protein), there is an approach that uses knockout animals, in which a gene is deleted or inactivated, as a method of analyzing the change with the loss of the gene. However, since technology and time are needed to produce a knockout animal that has lost a specific gene from genomic DNA, it is difficult to carry out analysis efficiently. Moreover, even when a knockout animal has been created, other genes may compensate for the inoperative gene, so all those genes may also have to be knocked out. It is very difficult to make knockout mice lacking the genes that are necessary for generating processes and cell functions. Techniques have been developed using antisense DNA, ribozyme, chimera oligo, etc., as a method of stopping the function of a gene more easily than knockout. The method and experimental operation of finally controlling protein expression by preventing the step of mRNA translation are simple. To design the DNA arrangement for an inhibition effect to be shown to acquire sufficient effect, many exploratory experiments need to be conducted.

RNA interference (RNAi) has appeared in recent years to solve the problems of the existing methods of gene control. Gene expression is controlled by introducing double-stranded RNA of 21–25 bases, making it possible to analyze gene function efficiently. Since control of the gene function by RNAi only controls gene expression, rather than destroying genomic DNA (knockout), this is called knockdown. This simplicity is establishing RNAi as an indispensable tool in molecular biology research, and both *in vitro* and *in vivo* analysis is increasingly being carried out (Guan et al., 2010; Tiscornia et al., 2003).

REGENERATIVE MEDICINE AND INDUCED PLURIPOTENT STEM CELLS

In 1996, the first mammal to be cloned from adult stem cells, "Dolly" the sheep, was born. Even though Dolly's birth was a major breakthrough, there were misunderstandings and opposition, because it became theoretically possible to create human clones using the same technique. In 2006, Professor Shinya Yamanaka's group at Kyoto University, Japan, converted adult cells into induced pluripotent stem cells (iPSCs). iPSCs are the same as embryonic stem cells (ESCs) in that they can grow into every type of cell in the body and can self-proliferate indefinitely. The difference between iPSCs and ESCs lies in how they are made. To obtain ESCs the early embryo has to be destroyed. In contrast, iPSCs can be made by introducing a specific gene into a cell from the adult skin, for example. This difference removes the problems inherent in ESCs, which are linked with the definition of life. The Ministry of Education, Culture, Sports, Science and Technology of Japan regards an early human embryo as the germ of a life, and states that it should be treated with care and respect so that people's dignity is not invaded, and this guidance also relates to ESCs. For this reason, there are strict conditions on using embryos and ESCs for research purposes, and the scrutiny of countries and research institutions is required. However, such problems do not relate to iPSCs made with adult cells. With regard to regenerative medicine in the future, iPSCs will encounter fewer problems than ESCs. Regenerative medicine could use cultured cells to replace damaged cells and diseased internal organs.

Human ESCs, which can differentiate into any cell type, had been anticipated to lead to breakthroughs in regenerative medicine since their discovery in 1998. However, problems arose with regard to rejection by the patient's immune system as well as ethical problems. Although human embryo clones could be made from a patient's cell with cloning technology to obtain the ESCs, the procedure is technically difficult and raises ethical concerns. Producing ESCs that have genetic codes matching those in the patients was desired. iPSCs fulfilled this need without using cells from human embryo clones.

The transfection reagent TransIT-293 was used in a method for the gene expression of retroviral vectors in normal human dermal fibroblasts (NHDF-Ad) as follows.

RetroNectin was used for cotransfection into G3T-hi cells about three kinds of vectors according to iPSC guidance. Cloning of the gene was carried out on a laboratory dish or plate with pE-Ampho vector, and pGP efficient vector and a quantity of revelation transgenics retroviral vector plasmid (pDON-5), a gene for iPSC guidance (OCT3/4, SOX2, KLF4 and LIN28). Since the three genes (SOX2, LIN28 and NANOG) are connected by the IRES sequence, it was transferred as single mRNA, and SOX2 and NANOG were then translated through IRES, independent of LIN28 (or OCT3/4).

Method
G3T-hi cells (2×10^6) were seeded on a 6 cm collagen–coated dish and incubated for 24 hours. The retrovirus transfection reagent TransIT-293 (10 µl) was used. The pE-Ampho vector (1 µg) and pGP vector (2 µg), which are the components of Packaging Kit Ampho, were cotransfected into 3HI cells using TransIT-293, and the gene expression retroviral vector NHDF-Ad RetroNectin was used for iPSC guidance.

Fixation
Cloning of the gene was carried out on a laboratory dish or plate. pE-Ampho vector, pGP efficient vector and a quantity of revelation transgenics retroviral vector plasmid (pDON-5) were used. G3T-hi cells (2×10^6) were

seeded on a 6 cm collagen-coated dish. On the next day, evaluation of the pE-ampho vector (1 μg) and pGP vector (2 μg), which are components of Retrovirus Packaging Kit Ampho, and the iPSCs, was carried out using TransIT-293 (10 μl).

Method

The iPSC colony obtained by the RetroNectin method was taken up and cloned. The copy number of a provirus inserted into a genome was calculated by real-time PCR using the Provirus Copy Number Detection Primer Set and Cycleave PCR Core Kit. Total RNA was recovered using the Fast-Pure RNA Kit. The gene expression of the ESC marker was checked by RT-PCR. The iPSC clone 253G1, established by Yamanaka's group, introducing three genes as controls, was similarly analyzed (Nakagawa et al., 2008).

Result

When the insertion copy number to the genome was measured about the established iPSC clone (iPS-3), there were about 10.2 copies.

Method

An iPSC guidance experiment was conducted on a normal human dermal fibroblast by the RetroNectin method and the polybrene method. Transgenics was performed using the green fluorescent protein AcGFP1 revelation retroviral vector, and the red fluorescence protein mStrawberry revelation retroviral vector. Each virus solution was mixed with an equivalent amount of AcGFP1 and mStrawberry revelation retroviral vector solution, and then diluted to suitable concentrations. The percentage of the cell population that expresses AcGFP1 and mStrawberry cannot be compared by flow cytometry 5 days after cotransfection, whereas comparison of both sides of AcGFP1 can be measured. It was suggested that the combination of pDON-5 vector and RetroNectin induces effective iPSC formation even in the lower copy number inserted in the genomes. Moreover, when comparison of 11 kinds of ESC marker genes was checked by RT-PCR, expression of the marker genes was hardly detected in the origin fibroblast (NHDF-Ad). By contrast, expression of these genes was observed in the iPSC clone (iPS-3 and 253G1).

Result

The RetroNectin method found a higher rate of double fluorescence-positive cells than the polybrene method. The difference was so great that the virus concentration was low, and it was possible to carry out the procedure with a little virus solution using RetroNectin and NHDF (TRACP and ALP double stain). Although iPSCs hold great promise for regenerative medicine, problems remain with the clinical acceptance of iPSCs as a treatment method. The development of iPSCs has huge potential in basic biology, clinical applications and various fields of research in the near future.

REFERENCES

Guan C, Ye C, Yang X, Gao J. A review of current large-scale mouse knockout efforts. Genesis 2010;48:73–85.

JIS K0124:2002. General rules for high performance liquid chromatography. Japanese Standards Association; 2002.

Matsunaga T, Hara A, Komoda T. Functional change in the HDL particle by oxidative modification and its contribution to atherogenesis. In: Komoda T, editor. The HDL Handbook. 1st ed. New York: Academic Press; 2010. p. 220.

Nakagawa M, Koyanagi M, Tanabe K, Takahashi K, Ichisaka T, Aoi T, et al. Generation of induced pluripotent stem cells without Myc from mouse and human fibroblasts. Nat Biotechnol 2008;26:101–6.

Tiscornia G, Singer O, Ikawa M, Verma IM. A general method for gene knockdown in mice by using lentiviral vectors expressing small interfering RNA. Proc Natl Acad Sci USA 2003;100:1844–8.

Tomlinson MJ, Tomlinson S, Yang XB, Kirkham J. Cell separation: terminology and practical considerations. J Tissue Eng 2012;4. 2041731412472690.

Wilson ID, Plumb R, Granger J, Major H, Williams R, Lenz EM. HPLC-MS-based methods for the study of metabonomics. J Chromatogr B 2005;817:67–76.

CHAPTER 7

Concluding Remarks

This postscript is added in the hope that additional sections on vitamins, bones and the brain will be included in the second edition of this book. Interested authors with suitable scientific backgrounds are being solicited to collaborate with us on the next edition. For all enquiries, please contact Dr Komoda directly by e-mail (tkalp1lp@gmail.com).

Thank you,
Tsugikazu Komoda

Biochemistry for Medical Professionals. http://dx.doi.org/10.1016/B978-0-12-801918-4.00007-4

SUBJECT INDEX

Printed in the United States
By Bookmasters